System on Chip Interfaces for Low Power Design

System on Chip Interfaces for Low Power Design

Sanjeeb Mishra

Neeraj Kumar Singh

Vijayakrishnan Rousseau

AMSTERDAM • BOSTON • HEIDELBERG • LONDON
NEW YORK • OXFORD • PARIS • SAN DIEGO
SAN FRANCISCO • SINGAPORE • SYDNEY • TOKYO

Morgan Kaufmann is an imprint of Elsevier

Acquiring Editor: Todd Green
Editorial Project Manager: Lindsay Lawrence
Project Manager: Punithavathy Govindaradjane
Designer: Alan Studholme

Morgan Kaufmann is an imprint of Elsevier
225 Wyman Street, Waltham, MA 02451, USA

ISBN: 978-0-12-801630-5

British Library Cataloguing in Publication Data
A catalogue record for this book is available from the British Library

Library of Congress Cataloging-in-Publication Data
A catalog record for this book is available from the Library of Congress

For information on all MK publications
visit our website at www.mkp.com

Copyright Permissions

Designations used by companies to distinguish their products are often claimed as trademarks or registered trademarks. In all instances in which Morgan Kaufmann Publishers is aware of a claim, the product names appear in initial capital or all capital letters. Readers, however, should contact the appropriate companies for more complete information regarding trademarks and registration.

CSI-2 ^{SM}, D-PHY ^{SM}, and DSI ^{SM} are service marks, and SLIMbus® is a registered trademark, of MIPI Alliance, Inc. in the US and other countries. MIPI, MIPI Alliance, and the dotted rainbow arch and all related trademarks and trade names are the exclusive property of MIPI Alliance, Inc., and cannot be used without its express prior written permission.

Figures below are Copyright © 2005-2015 by MIPI Alliance, Inc. and used by permission. All rights reserved.

Figures 4.35, 4.36, 4.37, 4.38, 4.40, 4.41, 4.42, 4.43, 4.44, 4.45, 4.46, 4.47, 4.48, 4.48, 4.49, 4.50, 4.51, 4.52, 4.53, 4.54, 4.55, 5.12, 5.13, 5.14, 5.17, 5.18, 5.19, 5.20, 5.21, 5.55, 5.57, 5.58, 5.59, 5.60, 5.61, 5.62, and 5.63

Figure below is reprinted with permission granted by ANSI on behalf of INCITS to use material from INCITS 452-2009[R2014]. All copyrights remain in full effect. All rights reserved.

Figure 7.6

Figures below are reprinted with permission from the Video Electronics Standards Association, Copyright VESA, www.VESA.org. All rights reserved.

Tables 4.1, 4.2, 4.3, 4.4, 4.5 and 4.6
Figures 4.13, 4.14, 4.15, 4.17, 4.18, 4.19, 4.23, 4.26, 4.27, 4.28, 4.29, 4.30, 4.31, and 4.32

Figures below are reprinted with permission from High-Definition Multimedia Interface Version 1.3a, Copyright, HDMI 1.3a. All rights reserved.

Figures 4.9, 4.10, 4.11, and 4.12.
Tables 4.11 and 4.12

Figures below are reprinted with permission from PCI-SIG. All rights reserved.

Figures 5.42, 5.44, 5.46, 5.47, 5.52, and 5.54

Figures below are reprinted with permission from USB 3.0. Copyright Intel Corporation. All rights reserved.

Figures 5.26, 5.27, C.1, C.2, C.3, C.4, and C.5.

Figure below is reprinted with permission from Intel® LPC interface specification v1.1, copyright © 2002, Intel Corporation.

Figure 8.3

v

Dedicated to my late brother, Sandeep,
who passed away at such a young age.

—Neeraj Kumar Singh

Contents

Acknowledgments

We would like to express gratitude to the people who helped us through this book; some of them directly and many others indirectly. It's impossible to not risk missing someone, but we will attempt anyway.

First and foremost, we would like to acknowledge Balamurali Gouthaman for writing the sensor, security, and input/output interface chapters; and Kiran Math for his help on the storage section of the book.

We would like to thank Stuart Douglas and David Clark for their help in reviewing the concept, structure, and content of the book and arranging for publishing with Elsevier—David, your meticulous reviews helped the book significantly.

Thank you so much Todd Green, Lindsay Lawrence, Punitha Govindaradjane and all the Elsevier publishing team for the outstanding work, help, guidance, and support; you have gone the extra mile to make the book what it is.

We would like to thank Intel management, in particular Pramod Mali and Siddanagouda S., for the support and encouragement.

Above all, we thank our family and friends for their understanding, support, and for being continuous sources of encouragement.

SoC Design Fundamentals and Evolution

This chapter discusses various system design integration methodologies along with their advantages and disadvantages. The chapter also explains the motivation for current system designs to move from "system on board" designs toward "system on chip" (SoC) designs. In discussing the motivation for the move toward SoC design, the chapter also discusses the typical chip design flow tradeoffs as well as how they influence the design choices.

INTRODUCTION

A system is something that achieves a meaningful purpose. Like everything else, it depends on the context. A computer system will have hardware components (the actual machinery) and software components, which actually drive the hardware to achieve the purpose. For example, talking about a personal computer (also commonly known as a PC), all the electronics are hardware, and the operating system plus additional applications that you use are software.

However, in the context of this book, by a *system* we mean the hardware part of the system alone. Figure 1.1 shows a rough block diagram of a system. The system in the diagram consists of a processing unit along with the input/ output devices, memory, and storage components.

Typical system components

Roughly speaking, a typical system would have a processor to do the real processing, a memory component to store the data and code, some kind of input system to receive input, and a unit for output. In addition, we should have an interconnection network to connect the various components together so that they work in a coherent manner. It should be noted that based on the usage model and applicability of the system, the various components in the system may come in differing formats. For example, in a PC environment, keyboard, and mouse may form the input subsystem, whereas in a tablet system they may be replaced by a touch screen, and in a digital

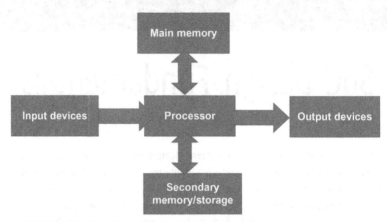

■ **FIGURE 1.1** A system with memory, processor, input/output, and interconnects.

health monitoring system the input system may be formed by a group of sensors. In addition to the bare essentials, there may be other subsystems like imaging, audio, and communication. In Chapter 3 we'll talk about various subsystems in general, involving:

1. Processor
2. Memory
3. Input and output
4. Interconnects
5. Domain-specific subsystems (camera, audio, communication, and so on)

Categorization of computer systems

Computer systems are broadly categorized as: general-purpose computer systems like personal computers, embedded systems like temperature control systems, and real-time systems like braking control systems. General-purpose computing systems are designed to be more flexible so that they can be used for different types of functions, whereas embedded systems are designed to address a specific function and are not meant to be generic. They are usually embedded as part of a larger device, and the user seldom directly interacts with such a system. Real-time systems are embedded systems with stringent response time requirements. All these computing systems are built using the same basic building blocks as shown in Figure 1.1. The flavor of the building blocks may vary from system to system because the design parameters and design requirements are different. For example, since embedded systems have a fixed known usage, the components can optimally be chosen to meet that functional requirement. The general-purpose system, on the other hand, might have to support a range of functionality and workloads, and therefore components need to be chosen keeping in mind the cost and user experience for the range of applications. Similarly the components for real-time systems need to be chosen such that they can meet the response time requirement.

SYSTEM APPROACH TO DESIGN

Due to the tighter budget on cost, power, and performance discussed in the previous section, the whole system is being thought about and designed as complete and not as an assembly of discrete pieces. The philosophy of system design thereby brings the opportunity to optimize the system for a particular usage. There is no real change in the system functionality; it's just that it is a different way of thinking about the system design. We already talked about the typical system components; next we will discuss the hardware software co-design, followed by various system design methodologies.

Hardware software co-design

As discussed earlier, a system in general has some hardware accompanied by some software to achieve the purpose. Generally, the system's functionality as a whole is specified by the bare definition of the system. However, what part of the system should be dedicated hardware and what should be software is a decision made by the system architect and designer. The process of looking at the system as a whole and making decisions as to what becomes hardware and what becomes a software component is called *hardware software co-design*. Typically there are three factors that influence the decision:

- Input, output, memory, and interconnects need to have hardware (electronics) to do the fundamental part expected from them. However, each of these blocks typically requires some processing; for example, touch data received from the input block needs to be processed to detect gestures, or the output data needs to be formatted specifically to suit the display. These processing parts, generally speaking, are part of the debate as to whether a dedicated hardware piece should do the processing or whether the general-purpose processor should be able to take care of the processing in part or full.
- The second factor that contributes to the decision is the experience that we want to deliver to the user. What this means is that, depending on the amount of data that needs to be processed, the quality of the output that is expected, the response time to the input, and so on, we have to decide the quality of the dedicated hardware that should be used,

and also this helps make the decision as to which processing should be done by dedicated hardware and which by software running on the CPU. The assumption here is that hardware dedicated to doing specific processing will be faster and more efficient, so wherever we need faster processing, we dedicate hardware to do the processing, such as, for example, graphics processing being processed by a graphics processing unit.

■ The third factor is the optimality. There are certain types of processing that take a lot more time and energy when done by general-purpose processing units as opposed to a specialized custom processor, such as digital signal processing and floating point computations, which have dedicated hardware (DSP unit and floating point unit, respectively) because they are optimally done in hardware.

System design methodologies

Early on, the scale of integration was low, and therefore to create a system it was necessary to put multiple chips, or integrated circuits (ICs), together. Today, with very-large-scale integration (VLSI), designing a system on a single chip is possible. So, just like any other stream, system design has evolved based on the technological possibilities of the generation. Despite the fact that system on a single chip is possible, however, there is no one design that fits all. In certain cases the design is so complex that it may not fit on a single chip. Why? Based on the transistor size (which is limited by the process technology) and size of the die (again limited by the process technology) there is a limited number of transistors that can be placed on a chip. If the functionality is complex and cannot be implemented in that limited number of transistors, the design has to be broken out into multiple chips. Also, there are other scalability and modularity reasons for not designing the whole system in one single chip. In the following section we'll discuss the three major system design approaches: system on board (SoB), system on chip (SoC), and system in a package (SiP) or on a package (SoP).

System on board

SoB stands for system on board. This is the earliest evolution of system design. Back in the 1970s and 1980s when a single chip could do only so much, the system was divided into multiple chips and all these chips were connected via external interconnect interfaces over a printed circuit board. SoB designs are still applicable today for large system designs and system designs in which disparate components need to be put together to work as a system.

Advantages of SoB

Despite the fact that this is the earliest approach to system design and back in the early days it was the only approach feasible to be able to do anything meaningful, the SoB design approach is prevalent even today and has a lot of advantages over other design approaches:

- It is quick and easy to do design space exploration with different components.
- Proven (prevalidated and used) components can be put together easily.
- Design complexity for individual chips is divided, so the risk of a bug is less.
- The debugging of issues between two components is easier because the external interfaces can be probed easily.
- Individual components can be designed, manufactured, and debugged separately.

Disadvantages of SoB

Since there is a move toward SiP/SoP and SoC, there must be some disadvantages to the classical SoB design approach; these can be summarized as follows:

- Because of long connectivity/interconnects, the system consumes more power and provides less performance when compared to SoC/SiP/SoP designs.
- Overall system cost is greater because of larger size, more materials required in manufacturing, higher integration cost, and so on.
- Since individual components are made and validated separately, they cannot be customized or optimized to a particular system requirement or design.

System on chip

By definition SoC means a complete system on a single chip with no auxiliary components outside it. The current trend today is that all the semiconductor companies are moving toward SoC designs by integrating more and more components of a system as SoC. However, there is not a single example of a pure SoC design.

Advantages of SoC

Some of the advantages of SoC design are

- lower system cost,
- compact system size,
- reduced system power consumption,

- increased system performance, and
- intellectual property blocks (IPs) used in the design can be customized and optimized.

Disadvantages of SoC

Even though it looks as though SoC design is very appealing, there are limitations, challenges, and reasons that not everything has moved to SoC. Some of the reasons are outlined below:

- For big designs, fitting the whole logic on a single chip may not be possible.
- Silicon manufacturing yield may not be as good because of the big die size required.
- There can be IP library/resource provider and legal issues.
- *Chip integration*: Components designed with different manufacturer processes need to be integrated and manufactured on one process technology.
- Chip design verification is a challenge because of the huge monolithic design.
- Chip validation is a challenge, also because of the monolithic design.

System in a package

SiP or SoP design is a practical alternative to counter the challenges posed by the SoC approach. In this approach, various chips are manufactured separately; however, they are packaged in such a way that they are placed very closely. This is also called a *multichip module* (MCM) or *multichip package* (MCP). This is a kind of middle ground between SoB and SoC design methodologies.

Advantages of SiP

In this approach the chips are placed close enough to give compact size, reduced system power consumption, and increased system performance. In addition:

- IPs based on different manufacturing technologies can be manufactured on their own technologies and packaged as a system.
- Because of smaller sizes of individual chips, the manufacturing yield is better.
- Development complexity is less because of division of design into multiple parts.
- Big designs that cannot be manufactured as a single chip can be made as a SiP/SoP.

Disadvantages of SiP

Despite the fact that the different chips are placed very closely to minimize the transmission latency, the SiP design is less than optimal in terms of power and performance efficiency when compared to SoC designs. In addition, the packaging technology for the MCM/MCP system is more complex and more costly.

In most of the literature, *SiP* and *SoP* are used interchangeably; however, sometimes they have different meanings. SiP refers to vertical stacking of multiple chips in a package and SoP refers to planar placement of more than one chip in a package. For example, a SiP or SoP can contain multiple components like processor, main memory, flash memory along with the interconnects, and auxiliary components like resistor/capacitor on the same substrate to make it a SiP or SoP.

Application-specific integrated circuit

Application-specific integrated circuit (ASIC) is a functional block that does a specific job and is not supposed to be a general-purpose processing unit. ASIC designs are customized for a specific purpose or functionality and therefore yields much better performance when compared to general-purpose processing units. ASIC designs are not a competing design methodology to SoC, but rather complementary. So, when designing an SoC, the designer makes a decision as to what IPs or functional blocks to integrate. And that decision comes based on whether the SoC is meant to be general purpose, catering to various different application needs (like a tablet SoC that can be used with different operating systems and then customized to serve as router or digital TV or GPS system), or for a specific purpose that is supposed to cater to only a specific application (e.g., a GPS navigator).

Advantages of ASIC

So, one might think that it is always better to make a general-purpose SoC, which can cater to more than just one application. However, there are significant reasons to choose to make an ASIC over a general-purpose SoC:

- *Cost*: When we make a general-purpose SoC and it is customized for a specific purpose, a good piece of logic is wasted because it is not used for that specific application. In case of ASIC, the system or SoC is made to suit; there is no redundant functionality. And therefore the die area of the system is smaller.
- *Validation*: Validation of an ASIC is much easier than the general-purpose SoC. Why? Because when a vendor creates a general-purpose SoC and markets it as such, there are an infinite number of possibilities

for which that SoC can be used, and therefore the vendor needs to validate to its specification to perfection. On the other hand, when one creates an ASIC, that piece is supposed to be used for that specific purpose. Therefore, the vendor can live with validation of the ASIC for that targeted application.

- *Optimization*: Since it's known that the ASIC will be used for the specific application, the design choices can be made more intelligently and optimally; for example, how much memory, how much should be memory throughput, how much of processing power is needed, and so on.

Disadvantages of ASIC

There are always tradeoffs. Of course, there are some disadvantages to the ASIC design approach:

- We all know that the hardware design and the manufacturing cycle are long and intensive (effort and cost). So, making an ASIC for every possible application is not going to be cost effective, unless we can guarantee that the volume of each such ASIC will be huge.
- Customers want one system to be able to do multiple things, rather than carrying one device for GPS, one for phone calls, another for Internet browsing, another one for entertainment (media playback), and yet another one for imaging. Also, since there are common function blocks in each of these systems, it is much cheaper to make one system to do it all, when compared with amortized cost of all the different systems, each dedicated for one functionality.

System on programmable chip

Because of a need for fast design space exploration, a new trend is fast gaining in popularity: the system on a programmable chip, or SoPC. In an SoPC solution there is an embedded processor with on-chip peripherals and memory along with lots of gates in a field-programmable gate array (FPGA). The FPGA can be programmed with the design logic to emulate, and the system behavior, or functionality, can be verified.

Advantage of SoPC

SoPC designs are reconfigurable and therefore can be used for prototyping and validating the system. Bug fixes are much easier to make in this environment than in an SoC design, where in one needs to churn in another version of silicon to fix and verify a bug, which has a significant cost.

Disadvantage of SoPC

The SoPC design models the functionality in an FPGA, which is not as fast as real silicon would be. It is therefore best fit for the system prototyping and validation, and not really for the final product.

System design trends

As we see from the preceding discussion, there are many approaches to a system design: one more suitable for one scenario than other. It should, however, be noted that the SoC approach, wherever possible, brings many advantages to the design. And therefore, not surprisingly, the SoC approach is the trend. However, for various reasons a pure SoC in ideal terms is not possible for a real system. In fact, initially it was only possible to design smaller embedded devices as SoC due to the limited number of transistors on a chip. It is now possible to integrate even a general-purpose computing device onto a single chip because Moore's law has allowed more transistors on a single chip. SoCs for general-purpose computing devices like tablets, netbooks, ultrabooks, and smartphones are possible these days. Given the advantages of SoC design, the level of integration in a chip is going to decide the fate of one corporation versus another.

HARDWARE IC DESIGN FUNDAMENTALS

In the previous section we talked about various system design approaches and the concept of hardware software co-design. Irrespective of the system design methodology, the computer system is made of ICs. We all know that the integrated chip design is a complex pipeline of process culminating in an IC chip that comes out of manufacturing. In this section we talk a little bit about the pipeline of processes in an IC design.

The basic building block of any IC is a transistor, and multiple transistors are put and connected together in a specific way to implement the behavior that we want from the system. Since the advent of transistors just a few decades back, the size of transistors has gone down exponentially, and therefore the number of transistors integrated in a chip has grown similarly. Just to bring in some perspective, the number of transistors on a chip in 1966 was about 10 as compared to billions of transistors on the latest one in 2014.

The minimum width of the transistor is defined by the manufacturing process technology. For academic purposes, the level of integration has been classified based on its evolution:

1. SSI = small-scale integration (up to 10 gates)
2. MSI = medium-scale integration (up to 1000 gates)

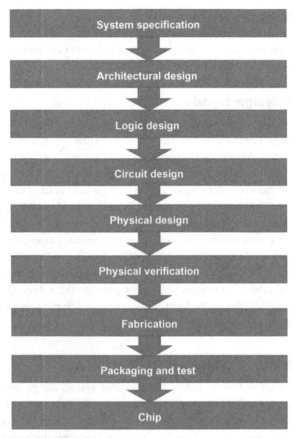

■ **FIGURE 1.2** High-level flow of chip design.

3. LSI = large-scale integration (up to 10,000 gates)
4. VLSI = very-large-scale integration (over 10,000 gates)

Given the complexity of the designs today, the IC design follows a very detailed and established process from specification to manufacturing the IC. Figure 1.2 illustrates the process.

CHIP DESIGN TRADEOFF

Tradeoff is the way to life. Tradeoff between cost and performance is fundamental to any system design. Cost of the silicon is a direct function of the area of the die being used, discounting the other one-time expenses in designing the IC. But changing set of the usage model and expectation from the computer system has brought in two other major design tradeoffs: that of power and that of configurability and modularity.

Power until a few years ago was a concern only for mobile devices. It is and was important for mobile devices because, with the small battery sizes required due to portability and other similar reasons, it is imperative that the power consumption for the functionality is optimal. However, as the hardware designs got more complex and number of systems in use in enterprises grew exponentially to handle the exponential growth in the workload, enterprises realized that the electricity bill (that's the running cost of the computer systems) was equally (or maybe more) important than one-time system cost. So the chip vendors started to quote power efficiency in terms of performance per watt. And the buyers will pay premium for power-efficient chips.

The other parameters configurability and modularity are gaining or rather have gained importance because of the incessant pursuit to shorten time to market (TTM). The amount of time it takes to design (and validate) a functional block in the chip from base is quite significant. However, if we look at the market, new products (or systems) are launched rather quickly. So, there is a need for the chip vendor to design a base product and be able to configure the same product to cater to various different market segments with varying constraints. The other factor that is becoming (and again in fact has become) important is modularity of the design. Why is modularity important? The reason again is that the TTM from conception to launch of a product is small, and the design of the functional blocks of system is really complex and time consuming. So, the system development companies are taking an approach to use the functional blocks from other designers (vendors) as IP and integrate that in their product. The fundamental requirement for such a stitching together is that the system design and IP design being sourced must both be modular so they can work with each other seamlessly. The approach helps both the system designer and IP designer: the system designer by reducing their TTM and the IP designer by allowing them to specialize and sell their IP to as many systems vendors as possible.

Chapter 2

Understanding Power Consumption Fundamentals

This chapter starts by explaining why power optimization is important, then tries to help the reader understand the sources of power consumption, measurement or how to monitor power consumption, and discusses the strategies applied to reduce power consumption at the individual IC and system level. However, before we start to delve into the details, a few things about why it's important.

WHY POWER OPTIMIZATION IS IMPORTANT

Saving energy is beneficial for the environment and also for the user. There is a lot of literature that discusses the benefits in detail, but to give just a few obvious examples, the benefits include lower electric bills for consumers, longer uptime of the devices when running on battery power, and sleeker mobile system design made possible by smaller batteries due to energy efficiency.

Knowing that power conservation is important, next we should discuss and understand the fundamentals of power consumption, its causes, and types. Once we understand them, we can better investigate ways to conserve power.

Since the use of electronic devices is prevalent across every aspect of our lives, reducing power consumption must start at the semiconductor level. The power-saving techniques that are designed in at the chip level have a far-reaching impact.

In the following section we will categorize power consumption in two ways: power consumption at the IC level and power consumption at the system level.

Power consumption in IC

Digital logic is made up of flip-flops and logic gates, which in turn are made up of transistors. The current drawn by these transistors results in the power being consumed.

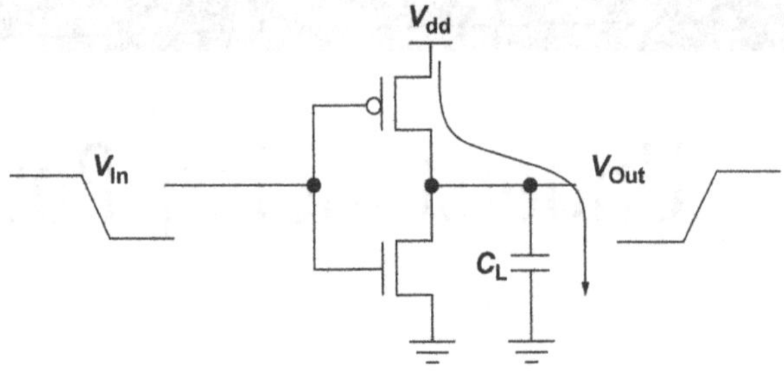

■ **FIGURE 2.1** Diagram of a transistor depicting voltage and current flow.

Figure 2.1 shows a transistor, voltage, and the current components involved while the transistor is functioning. So from the diagram, the energy required for the transition state will be $C_L*V_{dd}^2$. And the power (energy*frequency) consumption can be expressed as $C_L*V_{dd}^2*f$. Going further, the power consumed by the digital logic has two major components: static and dynamic.

Static power

Static power is the part of power consumption that is independent of activity. It constitutes leakage power and standby power. Leakage power is the power consumed by the transistor in off state due to reverse bias current. The other part of static power, standby power, is due to the constant current from V_{dd} to ground. In the following section we discuss leakage power and dynamic power.

Leakage power

When the transistors are in the off state they are ideally not supposed to draw any current. This is actually not the case: There is some amount of current drawn even in the off state due to reverse bias current in the source and drain diffusions, as well as the subthreshold current due to the inversion charge that exists at gate voltages under threshold voltage. All of this is collectively referred to as *leakage current*. This current is very small for a single transistor; however, within an IC there are millions to billions of transistors, so this current becomes significant at the IC level. The power dissipated due to this current is called leakage power. It is due to leakage current and depends primarily on the manufacturing process and technology with which the transistors are made. It does not depend on the frequency of operation of the flip-flops.

Standby power

Standby power consumption is due to standby current, which is DC current drawn continuously from positive supply voltage (V_{dd}) to ground.

Dynamic power

Dynamic power, due to dynamic current, depends on the frequency the transistor is operating on. Dynamic power is also the dominant part of total power consumption. Dynamic current again has two contributors:

1. Short circuit current, which is due to the DC path between the supplies during output transition.
2. The capacitance current, which flows to charge/discharge capacitive loads during logic changes.

The dominant source of power dissipation in complementary metal-oxide semiconductor (CMOS) circuits is the charging and discharging. The rate at which the capacitive load is charged and discharged during "logic level transitions" determines the dynamic power. As per the following equation, with the increase in frequency of operation the dynamic power increases, unlike the leakage power.

For CMOS logic gate, the dynamic power consumption can be expressed as:

$$P_{\text{dynamic}} = P_{\text{cap}} + P_{\text{transient}} = (C_L + C)\, V_{dd}^2\, fN^3$$

where C_L is the load capacitance (the capacitance due to load), C is the internal capacitance of the IC, f is the frequency of operation, and N is the number of bits that are switching.

So, fundamentally, as performance increases (meaning the speed and frequency of the IC increases) the amount of dynamic power also increases. It can also be noted that dynamic power is data dependent and is closely tied to the number of transistors that change states.

As is evident from the equation for power consumption, at the IC level, we have a few factors to tweak in: voltage, frequency, capacitance, and the number of transitions, to control or reduce the power consumption.

Power optimization in IC

So, the task of power minimization or management can be defined as: minimizing power consumption in all modes of operation (both dynamic when active and static when idle/standby) without compromising on the performance when needed. As discussed previously there are a few factors that we need to tweak in to optimize/minimize the power consumption:

- *Voltage*. As is evident from the equation, lowering the supply voltage quickly brings down the total power consumption. So, why don't we bring the voltage down beyond a point? It's because we pay a speed penalty for supply voltage reduction, with delays drastically increasing as V_{dd} approaches the threshold voltage (V_t) of the devices. This tends to limit the useful range of V_{dd} to a minimum of two to three times V_t. The limit of how low the V_t can go is set by the requirement to set adequate noise margins and control the increase in subthreshold leakage currents. The optimum V_t must be determined based on the current gain of the CMOS gates at low supply voltage regime and control of the leakage currents.
- *Capacitance*. Let us now consider how to reduce physical capacitance. We recognize that capacitances can be kept at a minimum by using less logic, smaller devices, and fewer and shorter wires. Some of the techniques for reducing the active area include resource sharing, logic minimization, and gate sizing. As with voltage, however, we are not free to optimize capacitance independently. For example, reducing device sizes reduces physical capacitance, but it also reduces the current drive of the transistors, making the circuit operate more slowly. This loss in performance might prevent us from lowering V_{dd} as much as we might otherwise be able to do.
- *Switching activity*. If there is no switching in a circuit, then no dynamic power will be consumed. However, this could also mean that no computation occurs. Since the switching activity is so dependent on the input pattern, for a general-purpose processor, containing switching activity may not be realistic. So, we do not focus on this. It should definitely, however, be kept in mind while defining and designing protocols so that we minimize the switching activity to as little as possible for average case scenarios.

Applying the fundamentals discussed in the previous section, to address the challenge of reducing power, the semiconductor industry has adopted a multifaceted approach, attacking the problem on three fronts:

- *Reducing capacitance*. This can be achieved through process development such as silicon on insulator with partially or fully depleted wells, CMOS scaling to submicron device sizes, and advanced interconnect substrates such as multichip modules. Since this is dependent on the process technology being used, there is a limitation depending on the current process technology, and any advancement has its own pace of development.
- *Scaling the supply voltage capacitance*. This is again a process-dependent factor, and it also requires change in the auxiliary circuit

and components in use. However, more importantly, the signal-to-noise ratio should be proper so that the communication is not broken because of noise signals of comparable strength.

■ *Using power management strategies.* This is one area where the hardware designer can make a huge difference by effectively managing the static and dynamic power consumption. It should however be noted that the actual savings depend a lot on the usage scenario or the application of the system. Some examples of power management techniques are dynamic voltage and frequency scaling (DVFS), clock gating, and so forth, which will be discussed in a little detail in subsequent sections.

In the next section we discuss these strategies in some depth. The various parameters interfere with each other, and therefore they cannot be chosen independent of other parameters or vectors. For example, CMOS device scaling, supply voltage scaling, and choice of circuit architecture must be done carefully together in order to find an optimum power and performance.

It may seem that process scaling can help solve all the power consumption problems; however, we should note that leakage power of the process of smaller size is more than leakage power of the process of higher size. Especially with 45 nm and below, the leakage power is more because of increased electric field. To counter this problem, new materials were discovered and employed. Silicon dioxide has been used as a gate oxide material for decades. The table in Figure 2.2 compares various parameters of silicon on three different process technologies. In the table 90 nm is taken as reference or baseline. Please note that the table compares the parameters in terms of multiplier and the values are not absolute. It should also be noted that the values are a rough estimate because these parameters will be influenced by other factors as well. The key takeaway or the point to note from

Parameter/technology	90 nm	65 nm	45 nm
Node length (nm)	1x	0.7x	0.5x
Frequency (GHz)	1x	1.43x	2x
Integration capacity (BT)	1x	2x	4x
Voltage (V)	1x	0.85x	0.75x
V_{th} (V)	1x	0.85x	0.75x
Dynamic power (W)	1x	0.7x	0.5x
Dynamic power density (W/cm^2)	1x	1.4x	2x
Leakage power density (W/cm^2)	1x	2.5x	6.5x
Power density(W/cm^2)	1x	2x	4x

■ **FIGURE 2.2** Comparison of various parameters driven by process in 90, 65, and 45 nm.

the table is that the leakage power is growing faster than device length going down.

So, a number of power-saving mechanisms are applied across design, architecture, and process to save dynamic power:

- *Multiple V_{dd}*:
    - *Static voltage scaling.* In an SoC, different blocks can work on different voltages, and the lower the voltage of a block, the less power it is likely to consume; therefore, it is imperative to create multiple voltage domains. To support this, typically voltage regulators are used to create different supplies from one source supply. IPs operating on one particular voltage will be put in the respective voltage island. So, in Figure 2.3, for example, the IPs operating on 1.2 V will sit on Voltage Island-1, while the IPs operating on 1.8 V will sit on Voltage Island-3, and so on. In Figure 2.3, the CPU is shown to be placed in a 1.2-V island, graphics and touch in a 1.8-V island, audio in a 1.5 V, and eMMC in a 1.9-V island. These voltage levels and the separation are just for the sake of illustration. The exact number of rails/islands depends on the design, IPs being used, and the intended usage model of the system.
    - *DVFS.* DVFS is a technique used to optimize power consumption in differing workload scenarios. In other words, the IP is designed in such a way that it does not consume fixed power all the time; instead, the power consumption depends on the performance level the IP is operating on. So, in heavy workload scenarios, the IP will be operating on a higher performance mode and thereby consuming higher power, while in lighter workload scenarios, the IP will be operating on lower performance mode and thereby consuming lower power. To implement this, in the IP design,

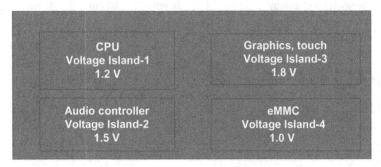

■ **FIGURE 2.3** System with multiple V_{dd}.

various performance modes are created. Each of the performance modes has an associated operating frequency and each of the operating frequencies has an associated voltage requirement. So, depending on the workload (and thereby the performance requirement), the system software can choose the operating mode. What this fundamentally means is that a particular IP is capable of running on multiple frequencies on respective voltages, and the software will choose the right frequency and voltage, at the point. A general design may look like Figure 2.4. As shown in the figure, the software will choose the right mode via mode control, which will translate to respective voltage and frequency setting.

■ *Adaptive voltage scaling (AVS).* A further extension to DVFS is AVS, wherein the mode controller monitors the state/performance requirement of the block and tunes the voltage/frequency of the block. In this design the need for software control goes away and therefore a finer control on DVFS is possible. A generic block diagram may look like Figure 2.5.

It must be noted that the mode monitor and controller unit continuously keep monitoring the IPs in different voltage islands and regulate the voltage based on the minimum required.

■ *Clock gating.* Since the clock tree consumes significant power (approximately 50% of dynamic power) it is important to reduce the power taken by the clock tree. Fundamentally clock gating means stopping the clock to a logic block when the operations of that block are not needed. This clock gating saves the power consumed by logic operating on each clock edge, power consumed by flops, and the clock tree itself. Many variations were devised that have built on this basic concept.

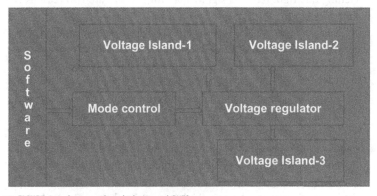

■ **FIGURE 2.4** System with multiple V_{dd} and DVFS.

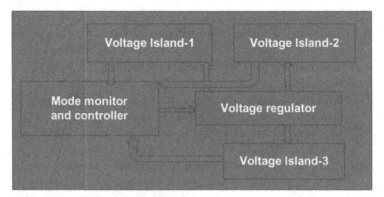

■ **FIGURE 2.5** System with adaptive voltage scaling.

■ *Frequency/voltage throttling.* This is one of the variations of clock gating, wherein the clock is not completely shut off, but rather, depending on the performance requirement of the clock frequency, is adjusted to a lower value such that the performance requirement is met with minimum power consumption. Since the supply voltage requirement depends on the clock frequency, the voltage is also adjusted appropriately to a low value, thereby reducing the power consumption even further.

■ *Power gating.* A logical extension to clock gating is power gating, in which the circuit blocks not in use are temporarily turned off. By power gating we bring the voltage to zero for devices not in use. For example, if the media IP is on a separate voltage rail, then it can be completely turned off when there is no media playback. This is made possible by the multiple voltage rails and domains in the design. Power gating saves the leakage power in addition to dynamic power. Getting this right requires significant effort. There are two reasons for this:

  – Since the time it takes to bring up the device from power off to power on is significant and noticeable, collaterals need to accommodate for power down state and define their operation flow accordingly.

  – The device may not be able to respond when powered down, and worse, may cause undesirable effects when accessed in power down state; the blocks accessing the powered-down units should include a mechanism to check whether the block can be accessed.

■ *Process improvement.* As transistors have decreased in size, the thickness of the silicon dioxide gate dielectric has steadily decreased to increase the gate capacitance and thereby drive current and raise device performance. As the thickness scales below 2 nm, leakage currents due to tunneling increase drastically, leading to high power consumption and reduced device reliability. Replacing the silicon dioxide gate dielectric with a high-κ material allows increased gate capacitance without the associated leakage effects.

So, to summarize the above discussion: There are various ways or mechanisms employed to save power, and these mechanisms do not work in isolation, but rather have interdependencies. Therefore various mechanisms are tweaked and put together to minimize or optimize the power consumption.

POWER CONSUMPTION OF A SYSTEM

Roughly speaking, systems have two modes when powered on, the first being the active mode when the system is actively being used, the second mode of operation being standby mode wherein the system is on but is on standby and waiting for input from the user. In the standby mode, to save power, most of the system components will be turned off since they are idle. To effectively manage the power and state transition, the Advanced Configuration and Power Interface (ACPI) standard defines various system states and device states in detail. Generally speaking, the device/IP is nonfunctional in low power states. In order to use the device/IP again, one needs to bring the device/IP back to a functional state from the low power nonfunctional state. The time taken in the process is called wake-up latency. Again, a general rule of thumb is, the lower the power state, the longer it takes to bring the device/IP to fully functional state (the more the wake-up latency).

So, speaking of the power consumed by a system, as shown in Figure 2.6, the total power consumed is a summation of active mode power consumption, standby (sleep) mode power consumption, and the wake-up power. In the figure the x-axis represents time, while the y-axis represents the power consumed at time x. Wake-up power represents the power wasted during wake-up. In a nutshell, there are three categories of power consumption, and separate strategies are applied to optimize each of them in a system:

1. Power consumption in active mode
2. Power consumption in standby mode
3. Power wastage during system wake

Power optimization at the system level

While discussing power optimization at the system level, we will discuss the optimization on the three fronts: active power management (APM), idle power management, and connected standby power management.

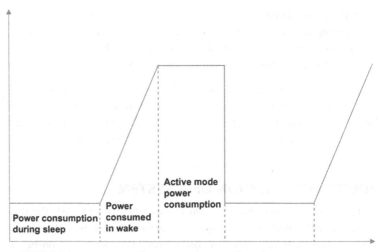

■ **FIGURE 2.6** Power consumption of system across active, standby, and transit.

Active power management

Active power management refers to the management of power when the system is being used. The main thing to understand about APM is that even when the system is in use, only a few of the subsystems are active; therefore the rest of the system components can be turned off. To this end, the system is being designed with use cases in mind, such that when a system is in use in a particular way, only the resources required for that use case are active and the rest can be power gated to save maximum power.

Idle power management

Idle power management is the set of policies that are employed to save the power when the system is idle. In modern-day systems, it is also desirable that the system is able to resume a normal full functional state as soon as there is need for it. The need may arise from an incoming call or the user's desire to wake the system for normal usage. The idle power management requires that the system is in a state where it consumes as little power as possible. However, the components are able to become functional in very little time. To this end, there is a lot of effort on the part of the system designers, hardware IP designers, and operating system (OS) designers.

Connected standby power management

Modern systems are not only supposed to be using little power when idle and come back up to working state when required, but there is a third dimension to it. That third dimension is that even when idle, the system is connected to the world and keeps up to date with all that is happening. For example, the system keeps the stock tab, news, and social media notifications all up to date so that when a user opens it up, the user finds everything up to date. In addition, the system should be able to notify the user of the events the user has subscribed to. To this end, the whole system is being designed in such a way that:

1. System components (at least some) have a state where they consume very little power, all the functional parts are shut down, but they have a portion that is always on and connected.
2. The entry and exit to the low power state is limited and predictable.
3. Offload system components have built-in intelligence such that it can function and do some basic jobs without involving other system components. For example, the network devices in a connected standby platform must be capable of *protocol offloads*. Specifically, the network device must be capable of offloading address resolution protocol, name solicitation, and several other Wi-Fi-specific protocols. And for another example, audio playback can be offloaded such that during audio playback only the audio controller is active and everybody else can go to low power states (after setting things up for the audio controller, of course).
4. Wake system components have a mechanism to wake the system when required. This occurs in three cases:
   - One of the offloaded components has discovered some event for which it needs to involve another system component.
   - One of the offloaded components needs the assistance of another component to carry out further instructions.
   - User has requested the system to come up for action via any of the interfaces (typically buttons).
5. The OS and software is designed in such a way that at every small interval the system comes up online, does routine housekeeping, updates the relevant tabs, and goes back to sleep. In this context, modern (OSs) have introduced a new concept, *time coalescing*, which simply means that all the recurring bookkeeping jobs are aligned such that the system is able to carry out all the tasks in one wake-up instance and they don't require a separate wake-up for each of them, which would be counterproductive to say the least.

ACPI states

In order to facilitate optimal power management at the system level, ACPI has defined standard states for system, devices, processors, and so on. Figure 2.7 shows the various states that are defined by ACPI and transitions

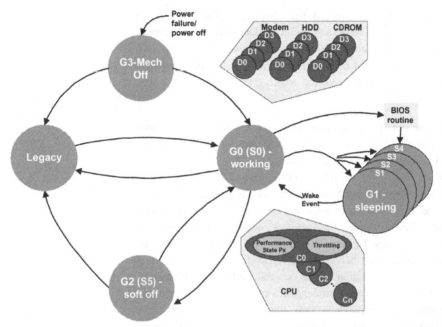

■ **FIGURE 2.7** Global system power states and transitions. *HDD*, Hard Disk drive; *BIOS*, Basic Input Output system. © *Unified EFI, all rights reserved, reprinted with permission from ACPI Specification 5.0.*

between them. In the following sections we talk about them and explain what they all mean.

Global and system states

ACPI defines four global states and a total of six system states. The global states are marked G0-G3 while the system states are marked as S0-S5. It must however be noted that even though S6 is mentioned in some motherboard documents, it is not an ACPI-defined state. S6, wherever mentioned, corresponds to G3.

ACPI defines a mechanism to transition the system between the working state (G0) and a sleeping state (G1) or the soft-off (G2) state. During transitions between the working and sleeping states, the context of the user's operating environment is maintained. ACPI defines the quality of the G1 sleeping state by defining the system attributes of four types of ACPI sleeping states (S1, S2, S3, and S4). Each sleeping state is defined to allow implementations that can trade off cost, power, and wake latencies.

1. *G0/S0*: In the G0 state, work is being performed by the OS/application software and the hardware. The CPU or any particular hardware device

could be in any one of the defined power states (C0-C3 or D0-D3); however, some work will be taking place in the system.

    a. *S0*: System is in fully working state.

2. *G1*: In the G1 state, the system is assumed to be doing no work. Prior to entering the G1 state, the operating system power management (OSPM) will place devices in a device power state compatible with the system sleeping state to be entered; if a device is enabled to wake the system, then OSPM will place these devices into the lowest Dx state from which the device supports wake.

    a. *S1*: The S1 state is defined as a low wake-latency sleeping state. In this state, the entire system context is preserved with the exception of CPU caches. Before entering S1, OSPM will flush the system caches.

    b. *S2*: The S2 state is defined as a low wake-latency sleep state. This state is similar to the S1 sleeping state where any context except for system memory may be lost. Additionally, control starts from the processor's reset vector after the wake event.

    c. *S3*: Commonly referred to as standby, sleep, or suspend to RAM. The S3 state is defined as a low wake-latency sleep state. From the software viewpoint, this state is functionally the same as the S2 state. The operational difference is that some power resources that may have been left ON in the S2 state may not be available to the S3 state. As such, some devices may be in a lower power state when the system is in S3 state than when the system is in the S2 state. Similarly, some device wake events can function in S2 but not S3.

    d. *S4*: Also known as hibernation or suspend to disk. The S4 sleeping state is the lowest-power, longest wake-latency sleeping state supported by ACPI. In order to reduce power to a minimum, it is assumed that the hardware platform has powered off all devices. Because this is a sleeping state, the platform context is maintained. Depending on how the transition into the S4 sleeping state occurs, the responsibility for maintaining system context changes between OSPM and Basic Input Output System (BIOS). To preserve context, in this state all content of the main memory is saved to nonvolatile memory such as a hard drive, and is powered down. The contents of RAM are restored on resume. All hardware is in the off state and maintains no context.

3. *G2/S5*: Also referred as soft off. In G2/S5 all hardware is in the off state and maintains no context. OSPM places the platform in the S5 soft-off state to achieve a logical off. *S5 state is not a sleeping state* (it is a G2 state) and no context is saved by OSPM or hardware, but power may still be applied to parts of the platform in this state and as such, it is not safe to

disassemble. Also from a hardware perspective, the S4 and S5 states are nearly identical. When initiated, the hardware will sequence the system to a state similar to the off state. The hardware has no responsibility for maintaining any system context (memory or input/output); however, it does allow a transition to the S0 state due to a power button press or a remote start.

4. *G3: Mechanical off*: Same as S5, additionally the power supply is isolated. The computer's power has been totally removed via a mechanical switch and no electrical current is running through the circuitry, so it can be worked on without damaging the hardware.

Device states

In addition to global and system states, ACPI defines various device states ranging from D0 to D3.

1. *D0*: This state is assumed to be the highest level of functionality and power consumption. The device is completely active and responsive, and is expected to remember all relevant contexts.
2. *D1*: The meaning of the D1 device state is defined by each device class. Many device classes may not define D1. In general, D1 is expected to save less power and preserve more device context than D2. Devices in D1 may cause the device to lose some context.
3. *D2*: The meaning of the D2 device state is defined by each device class. Many device classes may not define D2. In general, D2 is expected to save more power and preserve less device context than D1 or D0. Devices in D2 may cause the device to lose some context.
4. *D3 hot*: The meaning of the D3 hot state is defined by each device class. Devices in the D3 hot state are required to be software enumerable. In general, D3 hot is expected to save more power and optionally preserve device context. If device context is lost when this state is entered, the OS software will reinitialize the device when transitioning to D0.
5. *D3 cold*: Power has been fully removed from the device. The device context is lost when this state is entered, so the OS software will reinitialize the device when powering it back on. Since device context and power are lost, devices in this state do not decode their address lines. Devices in this state have the longest restore times.

Processor states

ACPI defines the power state of system processors while in the G0 working state as being either active (executing) or sleeping (not executing). Processor power states are designated C0, C1, C2, C3, ... Cn. The C0 power state is an active power state where the CPU executes instructions. The C1 through Cn power states are processor sleeping states where the processor consumes

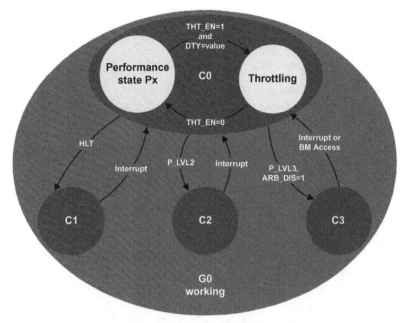

less power and dissipates less heat than when the processor in the C0 state. While in a sleeping state, the processor does not execute any instructions. Each processor sleeping state has a latency associated with entering and exiting that corresponds to the power savings. In general, the longer the entry/exit latency, the greater the power savings is for the state. To conserve power, OSPM places the processor into one of its supported sleeping states when idle. While in the C0 state, ACPI allows the performance of the processor to be altered through a defined "throttling" process and through transitions into multiple performance states (P-states). A diagram of processor power states is provided in Figure 2.8.

Now is the right time to ask that question: How do the low power interfaces reduce or optimize power consumption? The answer is simple: As we discussed earlier when introducing power consumption and strategies for power savings, the low power interfaces use the same fundamental mechanism in a way suitable for them to reduce power consumption; for example, idle detection and suspension or power gating/clock gating. In the forthcoming chapters we will discuss how these generic strategies are implemented in specific ways for specific interfaces/controllers/subsystems, based on the suitability. However, before we get there, we will discuss the functional aspect of various subsystems of a system in the very next chapter. That will be followed by the implementation details of each of the subsystem.

Chapter 3

Generic SoC Architecture Components

GENERIC SOC BLOCK DIAGRAM

As discussed in the previous chapters and illustrated in Figure 1.1, in Chapter 1, any computer system has input devices, output devices, processor, and memory. All the devices on the system are connected through interconnects. In today's world, computer systems are designed in a comprehensive manner; the computing power is distributed across the system and is not centralized at the CPU. Input and output devices are more intelligent, and connection interfaces are more scalable. So, to illustrate that, a real system block diagram is shown in Figure 3.1. The diagram shows the external view of Intel's Bay Trail platform designed for ultra-mobile devices like tablet and phones.

We now see the real instances of input and output devices and their connectivity. The real-world devices (like GPS, camera, and touch controller) are connected to the controllers implemented on SoC. We still have not talked about the specifics of interfaces connecting the real-world devices to the SoC platform. The Bay Trail platform is Intel's Intel Atom Z3000 SoC platform—this central piece integrates the controllers used for driving the real-world devices. Figure 3.2 shows the SoC's block diagram from the inside.

So, we see cores (CPU) connected with memory, storage, and other input/ output controllers like audio, graphics, camera, and low power input output (LPIO) via Intel On-Chip System Fabric (IOSF) bus or fabric and system agent. System agent is marked as PND SA in the block diagram. The *system agent* is the central arbiter and connectivity point that routes the transactions and requests from one controller to another.

To put the two diagrams together, the diagram in Figure 3.3 shows how the external components can be connected to the internal controllers. The diagram in Figure 3.3 shows one of the reference platforms. There could be different components, and they could be connected differently on a different reference platform.

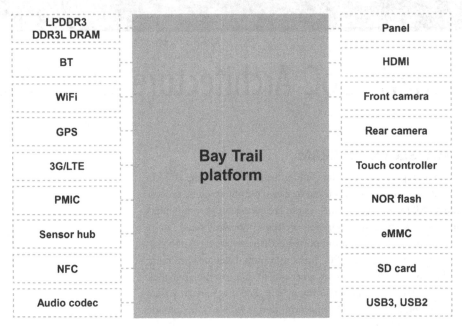

■ **FIGURE 3.1** Overview of the Bay Trail platform's connectivity.

SUBSYSTEMS OF AN SoC

Looking at the Bay Trail platform's connectivity diagram, there are plenty of components. These components can be further classified into subsystems, as will be described. It may, however, be prudent to mention that this classification is useful not only for logical grouping but also because these subsystem components have interdependencies among themselves, which means that the design choices for one affect the design choices of others.

Let's briefly go over the subsystem definitions, components that form a particular subsystem, and the design choices/dimensions of evaluation.

CPU

The CPU is the one fundamental component of the system. A number of vendors supply CPUs, and there are a number of factors affecting the decision on which one to choose. There are four key vectors used while choosing the CPU: instruction set architecture (ISA), ISA category, endianness, and performance.

Instruction set architecture

The CPU is the center of activity; it runs the operating system, which governs the functioning of the whole system. It should however be noted that the CPU can only carry out instructions in language it understands. The ISA

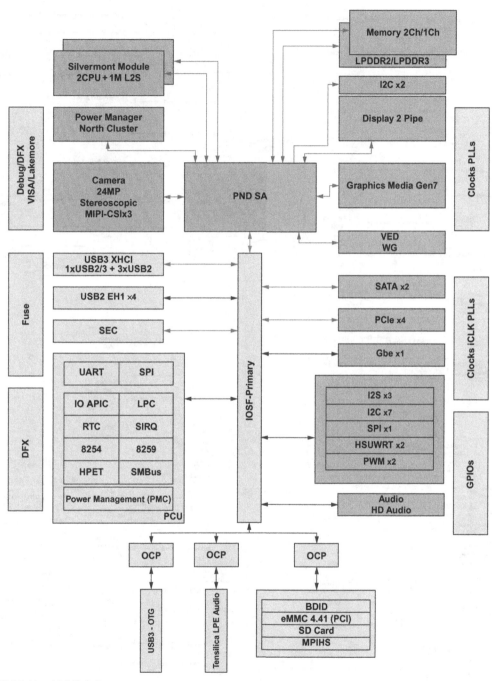

■ **FIGURE 3.2** Internal SoC block diagram.

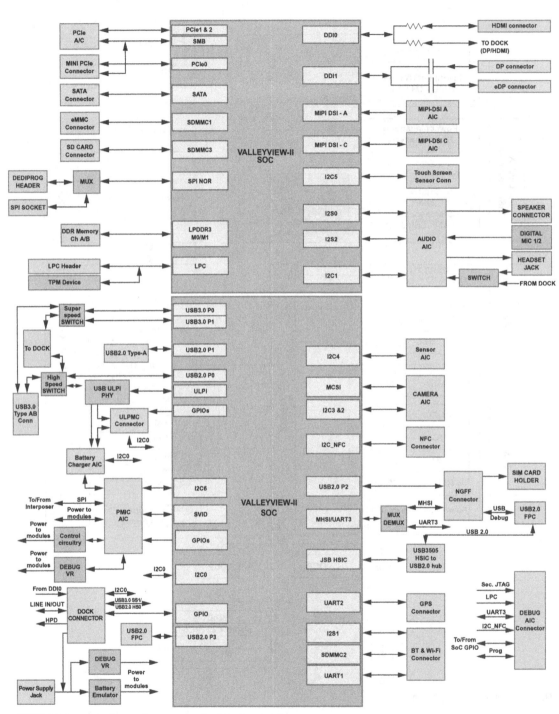

■ **FIGURE 3.3** Connectivity diagram of components on Bay Trail reference platform.

defines that language. The ISA is the part of the processor that is visible to the programmer or compiler writer. The ISA serves as the boundary between software and hardware. One can define one's own ISA and therefore the language that the hardware understands. However, since that is the bridge between the hardware and software, one needs to have software available for that architecture. Without appropriate software components there is no point having the hardware. Therefore, the first factor in choosing a CPU is: What is the ISA of the CPU we want?

There are a number of ISAs used across CPUs and microcontroller vendors. A few are popular across CPU hardware and software vendors: IBM PowerPC, Intel x86, and x64/AMD64; DEC Alpha, ARM ARMv1-ARMv8, and so on.

ISA category

It's more of an academic discussion, but all the ISAs have been roughly categorized between RISC and CISC. As the name suggests, RISC is reduced instruction set computer or CPU, while CISC is complex instruction set computer or CPU.

RISC generally has fewer instructions as part of the ISA, all of the instructions are the same size, and instructions are simple in nature. More complex operations are carried out using these simple instructions. The CISC is just the contrary: variable-length instructions, with many instructions supporting complex operations natively. It's easy to see that RISC and CISC have their own advantages; for example, RISC may be a lot more efficient from an instruction decode perspective, and simpler from a design perspective; however, CISC brings in value by potentially optimizing the implementation of the most frequently used complex instructions, as they implement them natively. To name a few prevalent RISC and CISC platforms:

- CISC: VAX, Intel x86, IBM 360/370, and so on
- RISC: MIPS, DEC Alpha, Sun Sparc, and IBM 801

There was a battle between RISC and CISC proponents, each claiming their own superiority. This battle lasted a long time, with proponents of each highlighting the advantages of their side and discounting the downside. The battle nearly came to an end when real commercial CISC implementations bridged the gap. The gap was filled by putting micro-sequencing logic in between the instruction decode and execution, as shown in Figure 3.4.

So, fundamentally, the execution units are RISC, but software thinks that it's CISC and supports all the various instructions—and the microcode running in between bridges the gap. This approach brings in the best of both worlds.

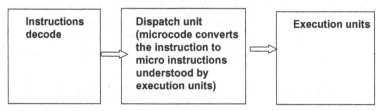

■ **FIGURE 3.4** CPU instruction decode and execute flow.

Endianness

Endianness relates to and defines how the bytes of a data word are arranged in the memory. There are two categories or classifications of endianness: little endian and big endian. *Big endian* systems are those systems in which the *most significant byte* of the word is stored in the *lower address* given, and the least significant byte is stored in the higher. Contrary to that, *little endian* systems are those in which the *least significant byte* is stored in the *lower address*, and the most significant byte is stored in the higher address. To illustrate the point with an example: Let us assume a value 0A 0B 0C 0D (a set of 4 bytes) is being written at memory addresses starting at X. For this example the arrangement of bytes for the two cases will be as illustrated in Figure 3.5.

Performance

Finally, performance is another major vector while choosing the CPU. For example, based on the usage of device, system designer will choose a high performance CPU vs low performance CPU or the other way around.

Bus, fabric, and interconnect

Bus, or fabric in the context of computer architecture, is a communication system that transfers data between components within a computer, or across computers. So fundamentally, it is a mechanism to interconnect various components and establish communication between them.

As part of implementation of communication across various components, there is a need to bridge throughput/speed gaps, clock speed deltas, and so on. So, for that sake, clock crossing units, buffers, and the like are deployed.

There has been a lot of advancement in the bus and fabric technology. Since bus and fabric are only the enablers, the reason for change in the bus and fabric technology is simple: scalability, modularity, and power efficiency. *Scalability* means the ability of the bus to deliver the higher throughput, *modularity* means the ease of putting different components together and making them talk to each other, and *power efficiency* of course means reducing power consumption as much as possible without sacrificing performance. The need for

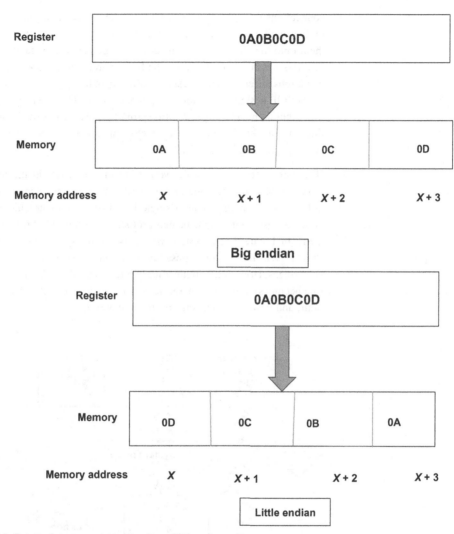

■ **FIGURE 3.5** Illustrating byte arrangement in big endian and little endian machine.

scalability and power efficiency is self-evident. However, the need for modularity is little less obvious. So, to elaborate a little bit on that, the need for modularity arises because SoCs typically integrate multiple IPs, and the majority of these intellectual properties (IPs) are designed by third parties who cater to various SoC developers. So, it becomes important for SoC designers to be able to integrate IPs from different vendors and make them talk. That's where the modularity of the SoC design becomes important.

Early SoCs used an interconnect paradigm inspired by the microprocessor systems of earlier days. In those systems, a backplane of parallel connections formed a *bus* into which all manner of cards could be plugged in. In a similar way, a designer of an early SoC could select IP blocks, place them onto the silicon, and connect them together with a standard on-chip bus. It is worth noting that since the IPs integrated in the SoC are delivered by various different vendors, standardization of the bus protocol was needed for both IP designer and SoC designers to work. The advanced microcontroller bus architecture specification provided the much-needed standardization and quickly became the de facto standard in the SoC world for IP development and integration.

However, buses do not scale well. With the rapid rise in the number of blocks to be connected and the increase in performance demands, today's SoC cannot be built around a single bus. Instead, complex hierarchies of buses are used with sophisticated protocols and multiple bridges between them. In Figure 3.6, a system with a two-level system bus hierarchy is shown. Note that it could possibly go to any level with more and more bus bridges. However, with the multiple levels of hierarchy, the timing closure becomes a problem. Therefore bus-based interconnects are reaching the limit, and newer mechanisms are being devised.

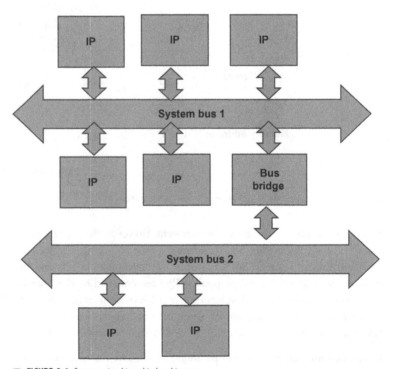

■ **FIGURE 3.6** System using hierarchical architecture.

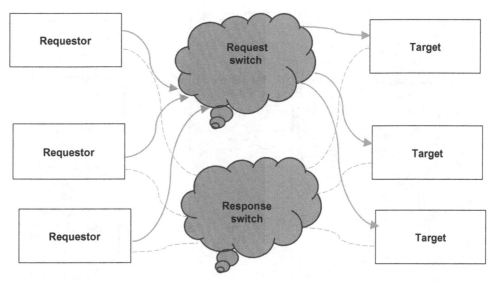

■ **FIGURE 3.7** Transaction flow in NoC-based designs.

To overcome the limits of bus-based solutions, packet-switched-network-based interconnects are being devised and applied. The architecture is termed *Network on Chip* or NoC. A packet-switched network offers flexibility in topology and tradeoffs in the allocation of resources to clients. Packetizing transports address, control, and data information on the same wires and allows transactions to be sent over smaller numbers of wires while maintaining high quality of service for each transmission. Additionally, distributing the interconnect logic throughout the chip rather than having bridges as chokepoints greatly simplifies floor planning of the most complex chips. NoC technology allows a tradeoff between throughput and physical wires at every point within the on-chip communication network. A generic diagram of NoC architecture is shown in Figure 3.7. As shown in the diagram, the requests flow between initiator and target via a switch. It must be noted that the same IP can play the role of initiator and target at different points of time. In the diagram, request packets (commands) are shown as solid arrowed line while response is shown with dashed line.

Display

Display is one of the most fundamental components of a system. It's not mandatory to have a display unit in a system; however, most of the systems have them in some form. The type of display used varies significantly depending on the usage scenarios. Some examples include blinking LEDs, seven segment display, text mode display, or fully featured graphics mode displays. In the context of the book, the displays used are fully featured graphics mode displays.

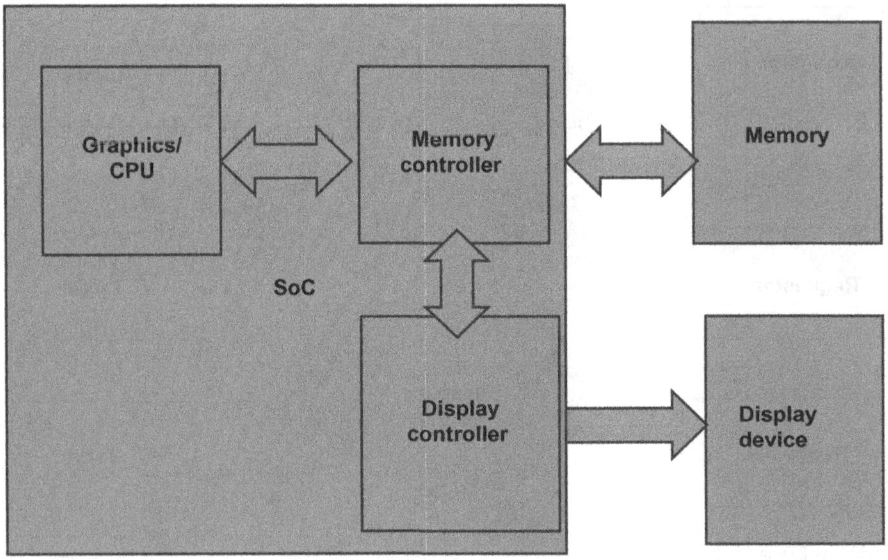

■ **FIGURE 3.8** Interaction and data flow between graphics and display controller.

The quality of display is measured in terms of number of pixels (a contracted short form for picture element). One pixel refers to one point on a display screen. Today's displays have very fine quality and can support high resolution. The advancements in display technologies and interfaces have been primarily driven by the need for higher quality and lower power. The relationship between graphics and display is shown in Figure 3.8: The CPU/graphics produce content to be displayed and write to memory, and the display controller fetches the content from memory and displays it on the display panel/device.

Multimedia

The multimedia subsystem encompasses the three major components of the system: audio, graphics, and imaging. These three components work so closely together that it is natural to view them as one subsystem. To illustrate the point, for example, a typical use case would be to record a video and be able to play it back. While recording the video, the imaging component captures the visual part, the audio component covers the audio capture, and while they are doing this, the graphics component is used for encoding functionality (meaning conversion of video from one format to another). Similarly, during playback, though imaging is not involved, the audio and graphics components work in tandem to make sure the playback is smooth and is in sync (also known as audio-video [AV] sync). Let's quickly discuss these three components and the trends in their areas.

Audio

By *audio* we refer to both the audio capture and playback components. By *audio playback*, we refer to playing back an already created or recorded audio media file. *Audio capture* refers to recoding the audio content. Audio interfaces have changed significantly from the inception, especially for the quality and usage scenarios. The audio content can be as raw as "a beep" to sophisticated premium content, also known as HD (high definition) audio. One system may support multiple audio interfaces to support different usage scenarios. Based on the usage scenarios the type and delivery mechanism is chosen. For example, early in the boot flow of a PC, beeps will convey messages: error or informational; however, while listening to music, the content may be played over a Bluetooth interface.

During audio playback and recording, the data-flow arrangement is like that shown in Figure 3.9. It is natural that at the time of playback the data will flow from processor to speaker via codec, while in case of recording, the data will flow from Mic to processor via codec.

It must be noted that the processor shown in the two scenarios in Figure 3.9 can be the CPU (the central processing unit) or a dedicated processor for audio processing. In Chapter 5, we'll discuss various interfaces used for audio in detail and will do a comparative study of them.

Graphics

By *graphics* we refer to the system components involved in the 3D rendering and media encode/decode/transcode. In practice, 3D rendering is used during gaming applications and normal user interface operations, whereas media encode and decode is used during video recording and playback.

To elaborate on 3D rendering: Needless to say, real-world scenes are 3D (three-dimensional: width, height, and depth). However, the majority of displays are 2D (two-dimensional: width and height). Though it is said that the majority of displays are 2D, in practice, almost all the displays in use today are 2D; only recently have displays with some 3D capabilities come to market, and they are not prevalent yet. So, fundamentally, what this means is that the real-world scenes' descriptions need to be processed and then converted to a form that can be displayed on 2D display and still convey the

■ **FIGURE 3.9** Data flow during audio playback and recording.

scene. This conversion is termed *3D rendering*. 3D rendering is a computationally intensive process. The finer the rendering, the more computationally intensive it becomes. Graphics devices have dedicated hardware to accelerate the processing. Because of the finer rendering requirements, newer generation graphics devices need to be more and more powerful. In practice, 3D rendering is applied during gaming, applications, during normal user interface operations, and so on.

The next thing that we want to talk about is media encode/decode/transcode. *Media* refers to content that has both video and related audio component. The next thing is *media encode/decode*: As we know, the media content has a lot of data. Why? Video content is made of a series of images, each known as a *frame*. For smooth video representation, 60 frames are captured for one second of video. As it has to keep data for each frame, you can sense the amount of data that needs to be stored for video content. However, readers would have realized by now that since we are capturing 60 frames or images in a second, there should not be a whole lot of difference between two consecutive frames. Therefore, to leverage this potential of optimization in size of recorded media content, specific encoding formats have been devised and applied. So, while recording or creating media content, the raw frame data is encoded in a specific format, and while playing back the encoded media content, the data is decoded and then played back. The first part refers to encode while the second part refers to decode. Pictorially, the same can be represented as shown in Figure 3.10.

The next thing that we need to discuss is transcode. As you might have guessed already, a number of encoding formats have been devised by different groups, teams, and organizations, each optimizing or trading off one parameter with another. And at times, one needs to convert content from one format to another. This process of converting content from one format to another is called *transcode*.

For a better user experience, the graphics performance should be high; because it has to process data real time, the graphics device has dedicated hardware to accelerate the 3D rendering, encode/decode, and transcode parts. Graphics devices are consequently becoming significant and are consuming an equally significant number of transistors/gates (chip area) on the system, requiring a lot of memory bandwidth.

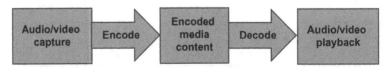

■ **FIGURE 3.10** Media encode/decode during audio/video capture and playback.

Since graphics devices have become more and more powerful, they process more and more data, and they require more and more memory bandwidth, the connection interfaces used to connect a graphics device to the system have changed. This is primarily because of the incessant bandwidth requirement for the huge amount of data that needs to be read, processed, and written back to the memory. In Chapter 4 we'll discuss the evolution and details of various buses used to connect graphics processors to systems, such as VL-bus, PCI, AGP, and PCIe.

Imaging

Imaging refers to the capturing of video content using a camera. The camera sensor captures the image data and passes it on to the processor for further processing. Again, the processor can be the CPU or a dedicated processor, typically called the image signal processor. Conceptually it looks very simple; however, imaging is now becoming more and more a differentiator in a product, so the imaging component supports a number of image quality features and supports high-resolution capture. Some of the quality features require continuous capture or multiple captures for each image. For example, when you click to capture an image, the imaging device may go ahead and do multiple captures and allow you to choose the best among them.

So, image quality and high-resolution capture features require higher bandwidth, and at the same time, the imaging devices are not active all the time, so there should be power management features that prevent the interfaces from consuming power when not in use. High bandwidth and power management requirements are the two key driving factors for new development in imaging interfaces. Chapter 5 discusses in detail various interfaces in use for imaging devices and provides a comparative study.

Communication

Because enabling communication is one of the fundamental requirements of mobile infotainment devices today, there are a lot of communication-related devices being put on the system, each of them enabling different usage scenarios. In the context of computer-based systems, communication can be categorized into wired and wireless. An example of wired communication is Ethernet, while BT, Wi-Fi, GSM, and GPS are examples of wireless communication. In the context of mobile infotainment devices, communication generally refers to mobile communication mechanisms.

Wireless communication is the transfer of information between two or more points that are not connected by an electrical conductor. The most common wireless technologies use radio as the carrier wave. This means the information signals are carried over radio waves. The process of mixing the information signal to the carrier wave is called *modulation*. The reverse process of extracting the information signal from the modulated signal is called *demodulation*. A communication device uses equipment called a modem (modulator/demodulator) for this purpose. A modem modulates the outgoing signal and demodulates the incoming signal. There are multiple methods for modulation. These include: amplitude and angle modulation as analog modulation methodologies, and phase-shift keying, frequency-shift keying, amplitude-shift keying, and quadrature amplitude modulation as digital modulation methodologies. Details of various modulation methodologies are beyond the scope of this book.

Different communication technologies like BT, Wi-Fi, cellular networks, and GPS use different carrier frequency ranges. The carrier frequency range (also known as frequency bands) is regulated by governments in most countries through a spectrum management process known as frequency allocation or spectrum allocation. Because radio propagation does not stop at national boundaries, governments have sought to harmonize the allocation of RF bands and their standardization. A number of standards bodies work on standards for frequency allocation, including:

- The International Telecommunication Union (ITU)
- The European Conference of Postal and Telecommunications Administrations (CEPT)
- The European Telecommunications Standards Institute (ETSI)
- The International Special Committee on Radio Interference (CISPR)

The choice of frequency band is based on the usage scenario, meaning, for example: Bluetooth is a short-range communication protocol, while GPS is a long-range communication protocol.

In the following sections, we quickly go over these usage scenarios and set the stage for Chapter 6, in which we discuss them in detail.

Bluetooth

Bluetooth is a short-range wireless communication technology standard. Bluetooth uses short-wavelength UHF radio waves of a frequency range between 2.4 and 2.485 GHz. Bluetooth enables one to create a personal area network wherein multiple devices talk to each other wirelessly via Bluetooth—a typical usage is home control automation systems. For example, the electronic devices in a home can be connected to a central control system via Bluetooth, and the central control system is controlled over the Internet. This usage scenario did not take off as anticipated; however, Bluetooth now is used to transfer data and control signal between two devices. These specific applications of Bluetooth are standardized as various profiles. There are many profiles defined, and two examples of these profiles are A2DP and hands-free profile (HFP). Advanced audio distribution profile (A2DP) is used for audio over Bluetooth, while HFP is used for hands-free operation of mobile phones. UART and USB are two of the leading interfaces for Bluetooth chips. UART is used when a Bluetooth chip is built into the system, such as in tablet devices. The USB interface is used when the Bluetooth module is connected as a separate dongle. We'll discuss the details in Chapter 6.

Wi-Fi/wireless LAN

Wi-Fi is a technology to enable electronic devices to communicate and/or connect to the Internet without wire. It uses 2.4-GHz UHF or 5-GHz SHF radio waves. Wi-Fi is a trademarked term owned by Wi-Fi alliance; it refers to implementation of IEEE 802.11 standard for WLAN.

Since by definition mobile infotainment SoCs are mobile in nature and they need to connect to the Internet for information/entertainment content, it's imperative that Wi-Fi has good throughput and consumes as little power as possible. Currently PCI and SDIO are two of the prevalent interfaces when the Wi-Fi module is connected as an integral part of the system. And, as usual, USB is the preferred interface, if and when the Wi-Fi module has to be connected externally. There is a push for an SDIO interface because of its lower power features, but many Wi-Fi chips still continue to use PCI. We'll discuss more details and provide a comparative study in Chapter 6.

Mobile telecommunication standards: 2G/3G/4G/LTE

Mobile communication is done over cellular networks. In fact, mobile communication started prior to the cellular system, and these systems were called 0G or mobile radio telephone. The 0G systems included push-to-talk (PTT), mobile telephone system, improved mobile telephone system, and advanced mobile telephone system. The cellular network system was enabled with the 1G standard, which used analog transmission by modulating the signal at a high frequency. The speed offered by 1G standard was 28-56 kbps. The 1G standard was replaced by digital 2G technologies. The first 2G technology, which was based on the GSM (Global System Mobile communications) standard, was originally developed by the ETSI. GSM used TDMA for multiplexing; however, there were other 2G technologies that used other multiplexing technologies. For example, CDMA-ONE uses CDMA, and PDC, iDEN, and D-AMPS all used TDMA. The GSM networks moved from 2G to 3G; however, in between there were interim enhancements that came as GPRS (General Packet Radio Service) and EDGE (Enhanced Data rates for GSM Evolution).

For applications like mobile TV, video conferencing, and mobile Internet, the data transfer rates needed to be much higher than those supported by 2G networks. To support that need, 3G wireless networks were defined. To be precise, a network has to be International Mobile Telecommunications-2000 (IMT-2000) compliant to be called 3G. CDMA2000 and UMTS are the two most prominent 3G networks.

The ever more demanding need for faster data required a successor to 3G even before it was fully deployed. The International Telecommunications Union-Radio communications sector (ITU-R) outlined the set of requirements for 4G standards, which was termed the International Mobile Telecommunications Advanced (IMT-Advanced) specification. The specification set peak speed requirements for 4G service at 100 Mbit/s for high-mobility communication (such as from trains and cars) and 1 Gb/s for low-mobility communication (such as pedestrians and stationary users). The first-release versions of Mobile WiMAX and LTE support much less than 1 Gb/s peak bit rate, so they are not fully IMT-Advanced specification compliant, but they are branded as 4G by service providers.

So, the ever-increasing need for higher data throughput and lower power yet again required the interface to be changed from one generation to another. In Chapter 6 we discuss in detail the various interfaces used and do a comparative study of the interfaces.

GPS

The global positioning system, better known by its acronym GPS, is a space-based satellite navigation system that provides location and time information. It needs unobstructed line of sight to four or more GPS satellites. In addition to GPS, other systems are in use. The Russian Global Navigation Satellite System (GLONASS) was developed in parallel with GPS; however, it suffered from incomplete coverage of the globe until recently. There are also the planned European Union Galileo positioning system, India's Indian Regional Navigational Satellite System, and the Chinese Compass navigation system.

All these navigation systems rely on the same philosophy, the need to receive and process the data from satellites in real time, and the need to make calculations on them to deduce the position parameters. Typically UART or USB is used for GPS interfacing. We discuss GPS in detail in Chapter 6.

FM

FM broadcasting is a broadcasting technology that uses frequency modulation over VHF (very high frequency) to provide high-fidelity sound over broadcast radio. The FM band here refers to the frequency range or band used for transmission or broadcasting of FM. It does not refer to the frequency modulation aspect of the transmission. The FM broadcast band falls usually between 87.5 and 108.0 MHz.

Near field communication

Near field communication (NFC) currently is used for contactless secure transactions, data exchange, device paring, and authentication. NFC is built on a set of standards that enable devices to establish secure radio communication with each other by touching each other, or when they are brought into proximity. NFC makes it easy for the technologies that need device paring or authentication.

Memory

Memory is one of the fundamental components of a system. There is at least some form of memory in a system. A number of technologies are used for making memory devices. However, in all, the memory devices can be classified into two categories: volatile and nonvolatile memory. Let's quickly review the two classifications and then we'll discuss them in detail in Chapter 7.

Volatile memory

Volatile memory is the memory that can keep the information only during the time it is powered up. In other words, volatile memory requires power to maintain the information.

Nonvolatile memory

Nonvolatile memory is the memory that can keep the information even when it is powered off. In other words, nonvolatile memory requires power while storing the data; however, once the data is stored, the nonvolatile memory technologies do not require power to maintain the data stored.

Volatile versus nonvolatile memory

As we can see, nonvolatile and volatile memory are fundamentally different by the definitions themselves. At first it may seem that nobody would prefer volatile memory over nonvolatile memory because the data are important and power is uncertain. However, there are a few reasons that both types of memories are in use and will continue to be in use:

- First and foremost, volatile memory is typically faster than nonvolatile memory, so typically when operating on the data it's faster to do it on volatile memory. And since power is available anyway while operating on or processing the data, it's not a concern.

■ Since, inherently, volatile memory loses data, the mechanism to retain data in volatile memory is to keep refreshing the data content. By refreshing, we mean to read the data and write it back in cycle. Since memory refresh consumes significant power, it cannot replace nonvolatile memory for practical purposes.

There is a memory hierarchy so that the systems can get the best of both worlds with limited compromises. A typical memory hierarchy in a computer system would look like Figure 3.11.

So, as depicted in Figure 3.11, the CPU continues to process data from nonvolatile memory, which is fast. However, the data in volatile memory is continuously backed by nonvolatile memory. It must be noted that if the memory CPU is talking to is slow, it would slow down the whole system irrespective of how fast the CPU is, because the CPU would be blocked by the data availability from the memory device. However, fast memory devices are quite costly. In practice, therefore, computer systems today have multiple layers in the memory hierarchy to alleviate the problem.

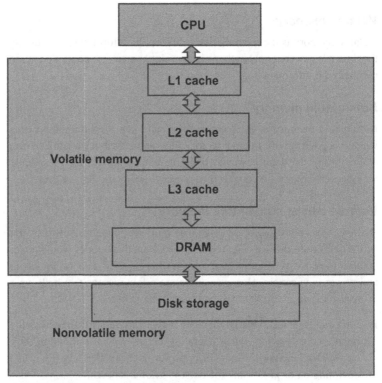

■ **FIGURE 3.11** Typical memory hierarchy of a computer system.

We can see that volatile memory has multiple layers in the hierarchy and typically the nonvolatile memory has a single layer. The layers in the memory hierarchy from bottom to top typically go faster, costlier, and smaller. The fundamental principle for having this multilayer hierarchy is called *locality of reference*. Locality of reference means that during a given small period of time, in general, data accesses will be in a predictable manner within an address region, and the switching in this locality will happen at intervals. Therefore the data in a locality can be transferred to the fastest memory so that the CPU can process the data quickly. This works not only in theory but in practice as well. Details of memory evolution and various interfaces that these memory devices use are discussed in Chapter 7.

Security

Computer systems today are used for e-commerce, and they contain sensitive information related to one's business, health, finance, and privacy. It is therefore imperative that the systems be secure and not able to be hacked into. Systems today employ security measures such that the system can be trusted for sensitive operations. The Trusted Computing Group (TCG) defines standards and references in the area of security and trusted computing. As per TCG: Trust is the expectation that a device will behave in a particular manner for a specific purpose. Therefore, any trusted platform should provide at least three fundamental features: protected capabilities, integrity measurement, and integrity reporting. Based on the three requirements, the Trusted Computing Group has defined the reference architecture for trusted platforms shown in Figure 3.12.

As shown in the reference architecture of design, the trusted platform module (TPM) module helps implement the three fundamental requirements a trusted platform needs to provide. In Chapter 8 we discuss in a little more detail how TPM helps implement the requirement and what interfaces are used to put a trusted platform module or device on the system.

Power

Needless to say, the system requires power for operations. And power is at a premium not just because it's a recurring cost manifesting in one's electricity bills, but also because portable devices are supposed to be mobile and run long hours without having to be recharged, so effective power delivery and management is critical to system performance. In this section we talk about three aspects of power: effective power delivery to the system; effective power management during active, standby, and idle states; and battery charging. For efficiency there is a separate IC, called the power management integrated circuit (PMIC), which is used for power-related activities. Figure 3.13 depicts the logical connection of PMIC, SoC, power supply, and charger.

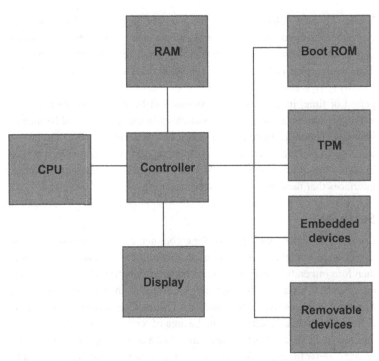

■ **FIGURE 3.12** Reference PC platform containing a TCG trusted platform module (TPM).

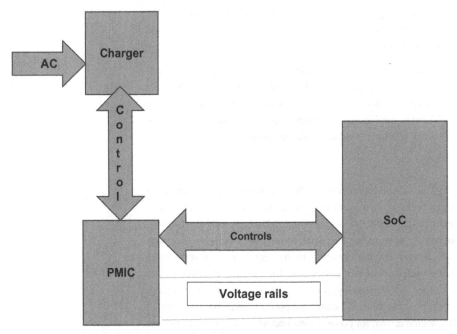

■ **FIGURE 3.13** Logical connection between PMIC, SoC, power supply, and charger.

Let's quickly discuss what these three aspects basically mean and then we'll discuss them in greater detail in Chapter 8.

Power delivery

Systems as SoCs have multiple components or IPs that run on different voltages. Each of these IPs or components is supplied power through separate rails. This brings in power efficiency because components or IPs that can run at lower voltage do not have to use higher voltage rails and lose unnecessary power in the system. The PMIC controls these rails.

Power management

The SoC and the PMIC issue a lot of control signals to indicate to each other specific states of the system. By doing so, the PMIC can save additional power by switching off the voltage rails that are not required to be on for that system state. The definition of various states is standardized so that there is no miscommunication between PMIC and SoC with respect to system state.

Battery charging

Mobile devices run on battery most of the time, and it needs to charge from time to time. Battery charging may appear to be the simplest of the parts in the system: One just needs to plug into AC power and the battery starts charging, right? That is actually only partially true. Because of the material used in making batteries and their sensitivity to temperature, and user experience issues, if overheated, the battery charging has to be closely regulated. Therefore, the charging has to be controlled actively such that conditions like excess heating and overcharge do not happen. It is due to this need that the battery charger is one of the most important parts of the system.

Sensor interfaces

Mobile infotainment devices are becoming smart by means of sensing lot of data around them and then using that data for various purposes. The sensing is accomplished by integrating sensing devices on the system. A *sensor* is a converter that can measure a physical quantity and convey that measurement through a signal. There are all kinds of sensors today. I am sure you'll find sensors for almost anything that you would want to sense. Search on the Web and you will be amazed to find what kinds of sensors are available. Just to give an example, there are sensors for sound, movement, chemicals, weather, flow, radiation, position, proximity, motion, force, pressure, and proximity, as well as optical, thermal, and electrical sensors.

Typically, the sensor is connected to some I/O interfaces that are used for transmitting data and control. Since the volume of data typically coming from sensors is not huge, low-speed, low-power interfaces are employed

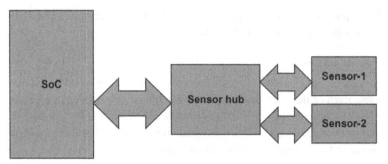

■ **FIGURE 3.14** System architecture with dedicated sensor processor as sensor hub.

to connect the sensors to the SoC. However, the number of sensors integrated into modern systems is growing significantly, and a lot of advanced and new features (like system wake on shake) are being implemented using sensors. These new features require the sensor data to be processed all the time, even when the system is in sleep (standby). It therefore made sense to have a discrete sensor processor. With that dedicated sensor processor the system architecture looks like Figure 3.14.

This architecture allows the system to offload the sensor processing work to a sensor hub and therefore the rest of the system (SoC) can go into lower power states.

Given that it was imperative to have a dedicated sensor hub, it made sense to integrate a sensor hub as part of the SoC. For this, the architecture would look like Figure 3.15.

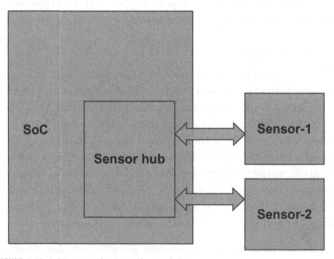

■ **FIGURE 3.15** Architecture with integrated sensor hub.

Integrating the sensor hub within the SoC brings the benefits of integration as part of the SoC along with the benefits of having a dedicated sensor processor. We'll discuss in more detail the interfaces and operations in Chapter 9.

Input devices

A system needs data to work on, to produce results, to take action based on those results, or just to display the output. The input can be implicit like sensor data, or explicit, which is provided by the user. In this section we talk about the input devices used for providing the inputs explicitly.

In the context of mobile infotainment devices, by input devices we refer to touch, keyboard, mouse, and remote control interfaces. In Chapter 10 we discuss in detail these various input mechanisms and the interfaces employed.

Debug interfaces

Due to the integration of multiple IPs on the system, an SoC is a complicated beast. It is therefore important for SoC designers to put in enough knobs so that any bugs can be analyzed, their root cause determined, and debugged effectively and efficiently. There are multiple interfaces used for debug purposes. Some of them are applicable for software debug, others for hardware debug. Also, some of them allow visibility into the system state and allow the debugger to make changes and see the results, while others only provide trace capability. There are so many kinds of debug interfaces that in Chapter 11 we will talk about the most popular one in greater detail.

CONCLUSION

In this chapter we quickly glanced over major components of a mobile infotainment device and provided a high-level overview. This sets the stage for detailed discussion of these components and the interfaces used for connecting them. In the following chapters we discuss each of the components one by one in a little more detail.

Chapter 4

Display Interfaces

Following up on the general discussion of various interfaces in Chapter 3, this chapter discusses, in detail, the various display interfaces, their applicability to various scenarios, and their capabilities. After this chapter, the reader should understand when to use a particular interface and what advantages a particular interface has over others for a particular design.

GENERIC DISPLAY CONCEPTS

Display is the primary output of a PC, tablet, or smartphone. It provides the visual output interface for the user. Every SoC has a *graphics processing unit*, commonly referred to as a GPU. The GPU is responsible for creating the image to be displayed by a process called *rendering*. The output of rendering is written to the system memory in a buffer, as shown in Figure 4.1. This buffer is called frame buffer.

The frame buffer contains the color information for each pixel that needs to be displayed. The format in which this information is stored in memory is known as the *pixel format*. Each pixel can be represented in either RGB format or YUV format. RGB is the most commonly used format. How many bits are used for each color component will further create different types like 8 bpc (8 bits per color) resulting in 24 BPP (bits per pixel), 6 bpc (6 bits per color component) resulting in 18 BPP, 10 bpc (10 bits per color component) resulting in 30 BPP, or 12 bpc (12 bits per color component) resulting in 36 BPP. Apart from this pixel information the end display requires the timing information also in order to display the image correctly. The timing information refers to the display resolution and refresh rate. The display resolution defines the screen size of the display; it is represented by number of pixels in one horizontal line and the total number of horizontal lines. A resolution of 1024×768 represents 1024 pixels $\times$ 768 lines or 1 frame. Hence the total number of bytes required for the frame buffer in this case will be $1024 \times 768 \times (BPP/8)$. And how often it is sent will determine the refresh rate of display, which is usually 60 Hz. All these when put together define the video mode: 1024×768 at 60 Hz, where 1024 refers to the *horizontal active*, that is, the active number of pixels in the horizontal direction; 768

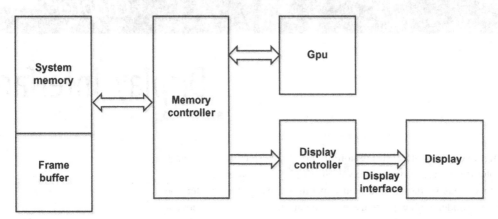

■ **FIGURE 4.1** Display data and control flow.

refers to the *vertical active*, that is, the active number of lines; and 60 refers to the refresh rate.

Apart from active pixels and lines there is a blanking interval giving rise to the horizontal and vertical blanking interval (see Figure 4.2). This blanking interval is not part of the physical display but only an analogy for representing time. The time delta that exists between two active lines is called the *horizontal blanking interval*. The time delta between two active frames is referred to as the *vertical blanking interval*. This time delta was needed in traditional cathode ray tube (CRT) displays. CRT displays have a phosphor screen and a cathode that emits an electron beam. When the electron beam comes in contact with the phosphor, different colors are created. The electron gun is moved across the entire phosphor screen creating the image for viewing. The electron gun moves from left to right covering one horizontal line, after which it needs to retrace to position itself accordingly. Also at the end of every frame a vertical retrace was needed to position the gun at the beginning of the first line. This gives rise to the horizontal and vertical totals:

$$H_{\text{Total}} = H_{\text{Active}} + H_{\text{Blank}}$$

and

$$V_{\text{Total}} = V_{\text{Active}} + V_{\text{Blank}}$$

Please note that all H^* use pixels as units, and all V^* use lines as the units. So the total pixel rate required for displaying a particular video mode is given by $H_{\text{Total}} \times V_{\text{Total}} \times \text{RR}$; this is often referred to as the pixel clock. For 1024×768 at 60 Hz, the H_{Total} and V_{Total} are 1344 and 806; $1344 \times 806 \times 60 = 65$ MHz. All this information is collectively known as *timing information*. Liquid crystal displays (LCDs), which do not have

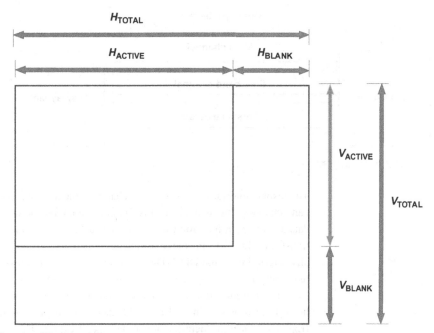

■ FIGURE 4.2 Illustration of blanking periods.

any electron gun, do not require the large blanking interval; hence the separate timing specification called reduced blanking timing (RBT) is defined for this purpose. The video mode 1024×768 at 60 Hz has an H_{Total} of 1184 and V_{Total} of 790, and with a pixel clock of 56 MHz as per RBT. So using RBT it is possible to support higher resolutions at lower pixel clocks.

The display controller within the SoC is responsible for transmitting both the color and timing information to the display using a particular display interface. There are different standards like Digital Visual Interface (DVI), High-Definition Multimedia Interface (HDMI), and DisplayPort (DP) for display interfaces; these are discussed in later sections. The display controller converts the color and timing information as per the required standard or protocol before transmitting the same. The rest of this chapter will describe the various types of display interfaces.

DISPLAY INTERFACE OVERVIEW

A display interface is an interface used for transmitting the color or pixel data and timing information from a source device to a sink device. A source device is a device that is the source of the pixel data and it is usually the SoC, and the sink device is the display (see Figure 4.3).

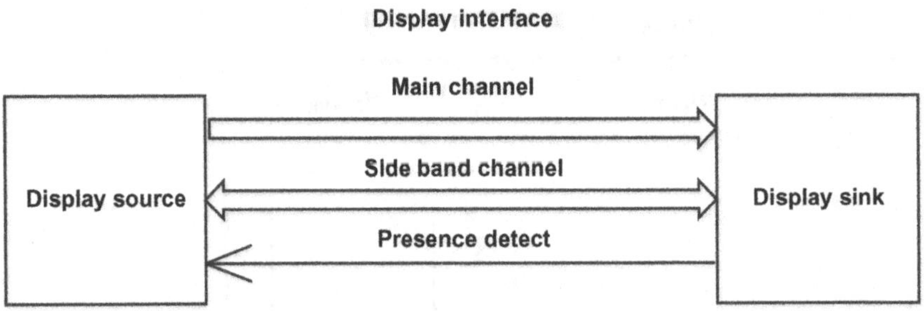

Display interface

Main channel

Side band channel

Presence detect

Display source

Display sink

■ **FIGURE 4.3** Display interface between source and sink.

The display interface consists of two channels: the main channel and a sideband channel. The main channel is a unidirectional bus on which both pixel data and timing information are sent from source to sink, and the direction of transfer is always from source to sink device. The sideband channel is usually referred to as the DDC (Display Data Channel), and it is used for two main purposes: display enumeration and monitor control. Each display can support different formats, video modes, and other capabilities. These capabilities are provided in an EEPROM inside the display as per the EDID (Extended Display Identification Data) specification. This data stored in the EEPROM can be read using the DDC by the source device. Apart from this, the display settings such as brightness and contrast that are normally adjusted using the buttons on the monitor panel can be controlled from the source. The Monitor Control Command Set (MCCS) is a specification meant for this usage. These are some of the usages of the sideband interface, and apart from this it is also used for handshaking during the authentication process in case of content protection. There are also some unique use cases for specific display interfaces that are discussed in later sections.

Extended Display Identification Data (EDID) and Display Data Channel (DDC)

As a number of different display panels with differing capabilities started to pour into the market, there was a need for a display source to identify the capabilities of a display panel in order for it to produce the best results. These capabilities include the maximum resolution that the panel supports and whether or not the display panel is capable of self-refresh.

The display panels/devices would store their capabilities in read-only memory (EEROM), which could be read by the display source, and the display source would act accordingly. In order to facilitate the requirement, two key pieces were needed: a standard way to define and store the capabilities and a standard mechanism/protocol for communication between the display source and display device. These two requirements led to the definition of two complementary standards. The first part covered standardizing capabilities and the structures used to document and store the capabilities, formally known as EDID. The second part covered the definition of the communication channel between the display source and display device, formally known as DDC.

Extended Display Identification Data

The Video Electronics Standards Association (VESA) defined the specification for EDID, and it contains information like manufacturer name, serial number, product type, phosphor or filter type, timings supported by the display, and display size. As per the EDID1.3 specification, the EDID data is organized as blocks of 128 bytes as shown in Table 4.1.

Each block is 128 bytes in size. The first block is mandatory, and the rest of the extension blocks are optional. If there is more than one extension block then block 1 is used as a block map. A block map describes each extension block as shown in Table 4.2.

Table 4.1 EDID Blocks.

Block#	Block Description
0	Base block
1	Extension
2	Extension
.	Extension
.	Extension
$N \leq 254$	Extension

Table 4.2 Block Map.

Byte#	Description
0	Tag for block map
1	Extension tag describing data in block 2 or block 129
2	Extension tag describing data in block 3 or block 130
.	
.	
127	Checksum

Table 4.3 Tag Description.

Tag	Description
0x01	LCD timings
0x02	Additional timing data type 2
0x20	EDID 2.0 extension
0x30	Color information type 0
0x40	DVI feature data
0x50	Touch screen data
0xF0	Block map
0xFF	Vendor-defined extension data

A byte of data is used for describing the extension data. Since there are only 128 bytes in any given block, block 1 cannot contain the tag information for all the extension blocks. Hence block 1 serves as a block map for extension blocks from block 1 through block 127, and block 128 serves as a block map for extension blocks from block 128 through block 254. Table 4.3 lists the tags assigned as per VESA.

Block 0 is the mandatory block as stated earlier, and the 128 bytes of this block are listed in Table 4.4.

Table 4.4 shows that there are three different ways in which the monitors can specify the supported video modes. The established timing field is 3 bytes (or 24 bits), listing a standard set of video modes for each bit. A monitor can set the appropriate bits to 1 based on which video modes it supports. The standard timing identification uses 2 bytes to describe a video mode, and it contains information like the number of horizontal pixels, aspect ratio, and refresh rate. The detailed timing descriptor uses 18 bytes to describe a video mode, and it contains information like pixel clock and the horizontal/vertical active and blanking regions.

Display Data Channel

As described in the previous section, the various information pertaining to a monitor's capabilities are stored in the form of EDID data structure and it is usually stored in read-only memory like EEPROM. The DDC is the physical interface, which the host uses to read this information from the monitor. The DDC uses the I^2C protocol to access the EDID information and also act as a command interface to control the monitor settings like brightness and

Table 4.4 EDID Block 0 Details.

Byte #	Number of Bytes	Field	Description
0x00	8	Header	0x00FFFFFFFFFFFF00
0x08	10	Vendor/product identification	Vendor name, serial number, etc.
0x12	2	EDID version/revision	Version and revision # in binary
0x14	5	Basic display parameters	Screen size in cm, supported features, etc.
0x19	10	Color characteristics	Colorimetry and white point information
0x23	3	Established timings	List of supported video modes
0x26	16	Standard timing identification	List of eight additional modes
0x36	18	Detailed timing descriptor #1	Lists the detailed timing information for a maximum of four modes or used for containing the monitor descriptors
0x48	18	Detailed timing descriptor #2 or monitor descriptor	
0x5A	18	Detailed timing descriptor #2 or monitor descriptor	
0x6C	18	Detailed timing descriptor #2 or monitor descriptor	
0x7E	1	Extension indicator	Number of optional extension blocks
0x7F	1	Checksum	The sum of all 128 bytes including the checksum should be zero

contrast. The DDC uses a point-to-point link between the host and the monitor using the two-wire I^2C protocol. The host is the I^2C master, and the monitor is the I^2C slave. Table 4.5 lists the pair of I^2C slave addresses used for DDC.

Using a pair of addresses, 256 bytes of information can be accessed. Using the Enhanced-DDC (E-DDC) addressing technique, up to 32 kB of data can be accessed. The entire 32 kB is divided into segments of 256 bytes each,

Table 4.5 I^2C Slave Address for DDC.

I^2C Slave Address	Purpose
0xA0/0xA1	EDID
0xA4/0xA5	EDID
0x6E/0x6F	DDC command interface for MCCS
0x60	Segment pointer address for E-DDC

Table 4.6 DDC Versions.

Name	Purpose	Comment
DDC	Display Data Channel	Generic term
DDC1	Original unidirectional mode	Not supported by VESA
DDC2	Original bidirectional mode	Not supported by VESA
DDC2B	A generic term for all bidirectional DDC modes	Not supported by VESA
DDC2Bi	Original name for command interface mode	Old terminology replaced by DDC/CI
E-DDC	Enhanced DDC, a bidirectional mode, supporting access to EDID data including all possible EDID extension blocks	Recommended by VESA
DDC-CI	The command interface uses I^2C single master communications and is used for MCCS	Recommended by VESA

and a segment pointer register is used to point to the segment that needs to be accessed. The segment point register is accessed using the I^2C slave address of 0x60 and is written with the appropriate segment value. After this the I^2C slave addresses 0xA0/A1 or 0xA4/A5 are used to read out the 256 bytes of the selected segment.

There are different versions of the DDC. Table 4.6 from VESA summarizes the different versions and what is supported now.

Presence detect

The presence detect, also referred to as Hot Plug Detect (HPD), is used by the source as a means to detect the plug and unplug events. Typically when a display device is plugged or unplugged the HPD is used to generate an interrupt to the CPU in the source. The device driver software takes the necessary actions based on this interrupt. Upon a plug event interrupt, the source will read the contents of the EDID and enumerate this newly attached display, and the user can choose to use this. Upon an unplug event interrupt, the source will remove this display from the list of enumerated displays, and the user cannot choose this any more.

DISPLAY INTERFACE CLASSIFICATION

The various display interfaces are Video Graphics Array (VGA), DVI, Low Voltage Differential Signaling (LVDS), HDMI, DP, embedded DisplayPort (eDP), and Mobile Industry Processor Interface (MIPI). These interfaces can be classified based on different aspects. One of the main aspects is the type of display that can be connected using the interface. On any device like a tablet, an Ultrabook device, or a smartphone, there is always a display that is permanently attached to these devices that can be referred to as the embedded, internal, or primary display. And it is also possible to connect these devices to an additional display like a TV or panels using a cable, which can be referred to as the external or secondary display. The SoC within the device can support different types of interfaces for connecting to both embedded and external display. Display interfaces like LVDS, eDP, and MIPI DSI are used for transmitting the pixel and timing data from the SoC to the internal display whereas interfaces like VGA, DVI, HDMI, and DP are used for doing the same on an external display.

EXTERNAL DISPLAY INTERFACE

We are now ready to discuss details of various external display interfaces, their evolution, and applicability.

Video Graphics Array (VGA)

VGA is one of the first interfaces developed for connecting the PC to an external display. The initial display devices were based on the CRT; this interface was also developed with CRT technology in mind. The VGA connector has 15 pins.

The DDC is implemented using the I^2C protocol. The I^2C protocol is based on two signals: serial clock line (SCL) and serial data line (SDA). There are two pins on the connector dedicated for this purpose. The I^2C address value of 0xA0/A1 or 0xA4/A5 is used for reading the EIDID from the display, and the I^2C address value of 0x6E/0x6F is used for MCCS. The main channel, which is used for transmitting the pixel and video timing information, comprises the following signals:

1. HSync
2. VSync
3. Red
4. Green
5. Blue

The HSync is a signal that is asserted once every horizontal line during the horizontal blanking interval, and VSync is the signal that is asserted once every frame during the vertical blanking interval. The frequency of VSync is the same as the refresh rate of the video mode; that is, if the refresh rate is 60 Hz, then the frequency of the VSync signal is also 60 Hz. These two signals will be used to directly control the speed and movement of the electron gun in the case of a CRT-based display. The red, green, and blue signals are the analog representation of the color information of the pixels. The voltages on these three lines will vary the intensity with which the corresponding phosphors on the screen get illuminated, which in turn will determine the final color of the pixel. Normally a DAC (digital to analog converter) in the SoC converts the color information stored in the form of digital bits to this analog signal for VGA. The peak-to-peak voltage of this signal is 0.7 Vpp.

Digital Visual Interface (DVI)

With the advent of LCD flat panels in the industry there was a need for defining a new interface, because the VGA display interface was developed specifically for CRT-based displays. The signals in the VGA interface directly controlled the electron gun of a CRT display. This was not needed any more for an LCD monitor because there was no electron gun in an LCD display. Furthermore, the representation of color information using analog signals is prone to errors, especially when it is transmitted over a meter-long cable. In order to overcome these limitations, it is best for the source to transmit the video information as digital data and the sink to convert the digital data to appropriate analog data based on the display technology. Since the digital to analog conversion happens finally within the display, this will be less error prone, which in turn improves the quality of the image being displayed. This resulted in the definition of DVI specification by the Digital Display Working Group (DDWG). DVI is the first digital display interface used for transmitting digital video data from a source such as PC to monitor. It is a very common interface for a digital flat-panel display.

Parallel digital display interface

Before getting to the details of the DVI specifications, let us first analyze what minimal set of digital signals are needed to convey both the timing and pixel information to the sink so that the sink can convert to the analog format for the display technology. A simple display interface for transmitting both the pixel and timing information will comprise the following signals shown in Figure 4.4.

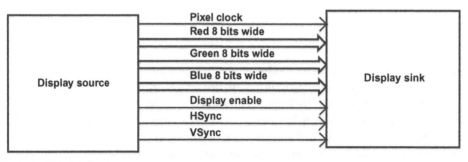

■ **FIGURE 4.4** Simple digital display interface.

The information that needs to be transmitted will include both color information and the video timing information. If 8 bits per color pixel format is assumed, then 24 bits will be required for conveying the color information for each pixel. Apart from this it will be required to send three other signals, which are display enable (DE), HSync, and VSync, hence a total of 27 bits is required. The DE signal will be high only during the transmission of pixels in the active region. The HSync and VSync bits will have a value of zero in the active region and in the blanking interval the HSync and VSync will have the appropriate values. The rate at which this information is transmitted should be the same as the pixel rate of the given video mode. It is therefore essential to transmit the pixel clock also, which will help the sink device to sample the other signals and reconstruct both the color and timing information.

So a simple display interface that will transmit the color and timing information to the display should have the following signals:

6. Pixel clock
7. HSync
8. VSync
9. Display enable
10. Twenty-four signals for color information, 8 bits for each color component.

Figure 4.5 shows how one active line of data is sent over such a display interface. In this diagram the DE is asserted for three clock cycles, which means

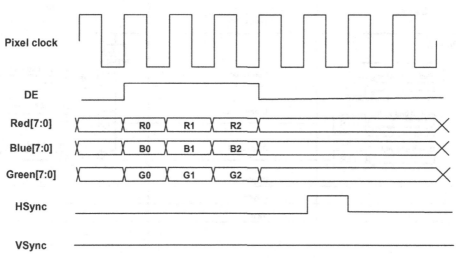

■ **FIGURE 4.5** Timing diagram: simple digital display interface.

the number of active pixels in one horizontal line in this example is three pixels. The color information for three pixels is sent during this time when DE is high. And HSync is high for one clock after the DE is low. With these sets of signals, the source can convey both the pixel information and the timing information to the sink. The display or sink will convert them to a suitable analog format based on the display technology. Such an interface can be used across all types of display technologies.

Serial digital display interface

This interface described so far works perfectly fine with the sink. It is very simple and eases the design of sink-side electronics. However, it is not possible to transmit these signals from source to sink through a cable. With 28 signals it is a challenge to deal with the electromagnetic interference on the signals, which can cause signal distortions especially at higher pixel clock rates. Common problems include crosstalk and ground noise. Reducing the number of signals will help contain these problems.

One way to reduce the number of signals is by serializing the data. As shown in Figure 4.6, the 8 bits of each color component can be serialized and sent on a single line. The source device incorporates a new functionality called a serializer (SER). The SER accepts 8 bits of parallel data and converts them to serial data. Within one pixel clock period 8 bits of data are transmitted serially on a single line. In effect the pixel clock period is subdivided into

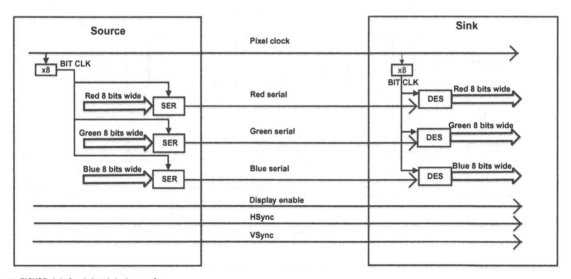

■ **FIGURE 4.6** Serial digital display interface.

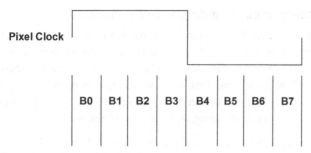

■ **FIGURE 4.7** Bit periods in serial digital display interface.

8-bit periods as shown in Figure 4.7. The bits are sent one after the other serially starting from bit 0 (B0) to bit 7 (B7).

In order to achieve this functionality, the SER needs to operate in a clock that is eight times faster than the pixel clock. This faster clock is commonly referred to as bit or serial clock. The sink device incorporates the converse of this functionality, which is nothing but a deserializer (DES). The DES receives serial data from the single input and constructs the 8-bit parallel data. It also uses the bit clock to sample the incoming serial data. This way the number of signals that are part of the display interface is greatly reduced to 7 from 28.

Differential signaling—Why it is important

Using a serial interface helps in reducing the number of signals and thereby prevents the interface from being bulky with many wires. However, transmitting data on a serial interface at a high speed and receiving them accurately is prone to errors. Using differential signaling, as opposed to single-ended signaling, makes the interface more immune to noise. Single-ended signaling uses a single ground as a reference for all the signals. The differential signaling technique on the other hand uses two complementary signals to represent the digital information. The information is represented as a voltage difference measured across the two wires.

This technique improves the immunity to the noise, since any noise will affect both the lines in the same manner and will finally get canceled at the receiver when the difference is calculated. The two signals are represented as p and n. One more problem remains to be solved, which is transitions. Whenever there are transitions (1 to 0 or 0 to 1) on the signal, this can lead to electromagnetic induction and can cause interference with other neighboring signals and lead to issues like crosstalk. Even with the use of a differential signal this can be a difficult problem to handle, especially at higher bit rates. The next section describes a method to reduce the transitions on the data.

Transition-minimized differential signaling

Reducing the transitions on the signal line greatly improves the quality of the signal at the receiver and makes the design of the receiver more robust. The transitions on the line can be reduced by transmitting the transitions instead of the actual bits. This is determined by comparing the current bit with whatever bit was transmitted previously. This type of comparison can be done very easily using an XOR or a XNOR function.

Let us analyze the truth table of a XOR function, shown in Table 4.7. If in Table 4.7 A represents the current bit and B represents a previously transmitted bit, then Y represents whether there is transition in the current bit with respect to the previous bit. How will this help reduce the transitions if Y was transmitted instead of A? A clearly noticeable advantage from the table is that both the 0 to1 and 1 to 0 transitions can be represented as 1. Hence, if there are many subsequent transitions with respect to the previous bit, then it will result in a continuous 1 being transmitted on the line, thereby reducing transitions. However, there is one drawback with this method when the subsequent bits are 1s. As per the last row in the table, if the previous and current bits are 1s then a 0 is transmitted. This introduces unwanted transition in the line when there are continuous 1s.

Let us also analyze the truth table of a XNOR function, as shown in Table 4.8. If in Table 4.8 A is the current bit and B is the previously transmitted bit, then

Table 4.7 XOR Truth Table

A	B	$Y = A$ XOR B
0	0	0
0	1	1
1	0	1
1	1	0

Table 4.8 XNOR Truth Table

A	B	$Y = A$ XNOR B
1	0	0
0	1	0
0	0	1
1	1	1

Y represents whether there is no transition in the current bit with respect to the previous bit. This is very similar to the XOR function. The XNOR has the same merit and demerit as the XOR function. In the case where there are many transitions it will result in 0 being transmitted, which will reduce the transitions, in the case where subsequent bits are 0s then a 1 is transmitted, introducing an unintentional transition. The limitation of both XOR and XNOR can be overcome by using the XOR function for all cases except the case when the number of 1s in the input stream is more than the number of 0s. In such situations the XNOR function can be used. This would mean that one more additional bit should be transmitted to the receiver to indicate whether XOR or XNOR function was used while transmitting the bits. And the receiver will analyze this additional bit first before reconstructing the actual bits from the received bits. The last issue that needs to be solved is maintaining the average DC value of the line at 0, which means the number of 1s and 0s transmitted in the line should be the same. This can be accomplished by keeping a count of 1s and 0s transmitted so far. By analyzing the counts it is possible to know whether more 0s or 1s have been sent so far (this is called *running disparity*). Based on this the data bits can be inverted to maintain the DC balance. This would mean that one more additional bit should be transmitted to receiver to indicate whether the data bits were inverted or not.

These two additional bits can be thought of as header bits that need to be sent along with the actual payload bits. How often these are sent will determine the overhead of transmitting these bits. These bits can be sent once every 8 bits. The incoming 8 bits of color information is encoded using XOR or XNOR. The ninth bit will indicate whether XOR or XNOR was used. The tenth bit indicates whether the first 8 bits are inverted or not. Hence, each 8 bits of data is encoded as 10-bit data. This 10-bit encoded data is serialized and sent on to the display interface on every pixel clock. With 10 bits it is possible to represent 1024 combinations, and not all of them will be used up for representing the 8-bit color information. The unused 10-bit codes can be used to represent the HSync and VSync values. The 2 bits of HSync and VSync will require four dedicated 10-bit values.

These unique codes are sent during the blanking interval and are referred to as *control codes*. And during the active region the values will be encoded as described earlier. Control data characters are designed such that they have a large number (seven) of transitions; this helps the receiver synchronize its clock with the transmitter clock.

The receiver has to implement the converse functionality to recover the 10-bit data from the incoming stream and decode the 10-bit data. If the 10-bit data is one of the four unique control codes, then they will be decoded accordingly. If

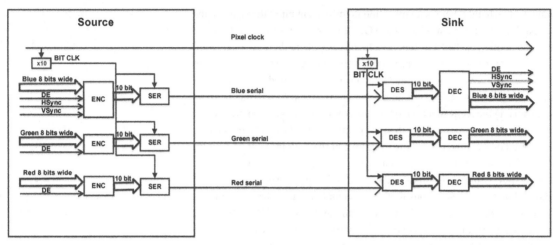

■ FIGURE 4.8 DVI display interface.

not, the ninth and tenth bit of the 10-bit code indicate what operation was done by the transmitter; based on this the receiver will decode the 10-bit data to get the 8-bit data. Figure 4.8 shows the details of the source and sink for such an interface, and this type of display interface is known as DVI.

The source has the same set of 27 digital signals as shown in Figure 4.5. The pixel clock is transmitted as a differential clock. Apart from this there are three other differential data signals, shown as Red Serial, Green Serial, and Blue Serial. During the active region, that is, when DE is asserted, the ENC (encoder) associated with each data line converts the 8 bits of color information to 10 bits using a transition-minimized differential signaling (TMDS) encoding scheme. During the Blanking Period the ENC of the Blue Serial line will convert the two control bits HSync and VSync to 10-bit control codes. The ENC of Green and Red Serial lines will send 10-bit control code corresponding to zero values of the two control bits. The 10 bits are serialized using the SER and sent on the respective differential serial signal. The SER operates on a 10x serial clock. The sink uses DES to de-serialize the incoming serial bits and it uses the 10x clock to recover the serial incoming bits from the data lines. The sink looks for the control codes to align to the 10-bit boundary. The decoder (DEC) of all the three data lines will decode the 10-bit codes to 8-bit data. The DEC associated with the Blue Serial line will also regenerate the DE, HSync, and VSync signals. The entire set of 27 signals is regenerated at the sink, which can be now converted to the appropriate analog format based on the display technology.

However, the numbers of signals required for carrying this information from source to sink are only four differential pairs.

DVI TMDS link architecture

DVI has two types of interface: single link and dual link. The single link interface has one clock pair, three pairs of data channels, and can support pixel rates up to 165 MHz.

The dual link interface has one clock pair, six pairs of data channels, and can support pixel rates higher than 165 MHz. The odd pixels are sent on one link (comprising three data channels) and the even pixels are sent on the other link (using the other three data channels). Since pixel information for two pixels are sent on each clock, it is sufficient to send the clock with half the frequency on the clock channel. For example, for a video timing requiring 200 MHz pixel rate, a clock with 100 MHz frequency is sent on the clock channel, and the odd pixels are sent on data channels 0, 1, and 2, and even pixels are sent on data channels 3, 4, and 5.

Apart from these signals, there are signals for DDC. As described earlier, this is a two-signal interface used for enumeration of the display and also for monitor control. Also there is a signal for hot plug detection. Whenever the external cable is plugged into or unplugged from the system, the operating system (OS) should be notified of this event to take the necessary actions. On a Hot Plug Event, an interrupt should be generated to enable the OS to initiate the display enumeration process. At the end of the enumeration it is possible to provide the information of the monitor that got attached to the end user's device. On a Hot Unplug Event, an interrupt should be generated so that the OS is aware that there is no display attached to this connector and the necessary action can be taken.

DVI was the first-ever digital display interface using only four differential pairs of signals; it was possible to send the pixel and timing information to the sink. It was designed such that it was generic enough to be used with any display technology. It served as a very good interface for PC monitors. However, the use of DVI in televisions in the living room was very limited.

High-Definition Multimedia Interface (HDMI)

Over a period of time it was possible for the various devices like laptops, tablets, and smartphones to play high-quality video. And at the same time advancement in TV technology made it possible to replace CRT-based TVs with LCD-based TVs in the living room. Providing the ability to connect one's device to TV and watch videos or look at photos on a bigger screen provided end users with a lot more options or usages for the end devices. They could record videos or take photos using a tablet or smartphone and watch them on TV. Also, they could buy premium content from the Internet and watch it on their TVs by connecting the device to the TV. This encourages the users to do more and more with a single device of their choice. Watching a video meant both audio and video were important. Using DVI, the high-quality video could be transmitted. However, a separate interface was needed for audio. Having multiple cables or a bulkier connector was not very attractive for consumers, both from an aesthetic and cost perspective. A connector and cable that could carry both the audio and video information on a cost-effective and small connector form factor was needed. Also DVI supported a maximum of 24 BPP; it did not support the other formats like 30 BPP, 36 BPP, and so on. There was a growing market need to support these formats for a better visual experience. These were some of the main driving factors for defining an additional specification after DVI, referred to as HDMI.

The HDMI block diagram is shown in Figure 4.9. It has one differential clock pair and three differential data pairs like DVI. The clock is referred to as the TMDS clock, whose frequency is either equal to or greater than

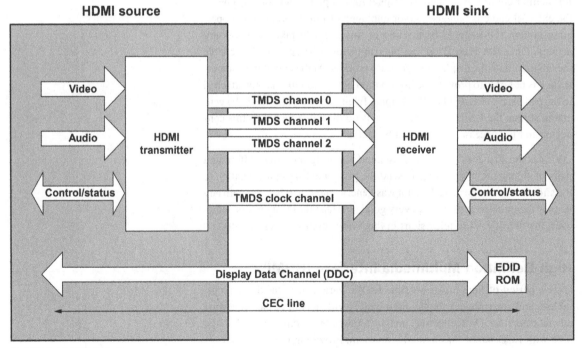

■ **FIGURE 4.9** HDMI block diagram.

the pixel clock depending on the pixel format. The three data pairs used for carrying video, audio, and other auxiliary data. The DDC is used for display enumeration and status exchange. There is an optional CEC (consumer electronic control) line, which is used for providing control functions to users for controlling their various consumer electronics products.

The video data is sent using TMDS encoding like DVI. The audio and auxiliary data is packetized and sent during the blanking period.

Data encoding and packetization

As stated earlier, the frame is divided into an active region and blanking region. During the active region, color information will be sent, whereas during the blanking region only 2 bits of sync information are sent. Quite a lot of bandwidth is left unused during the blanking region, which can be used for sending additional information. The additional information could be audio or other control information. This way it is possible to send both audio and video to the TV (display) using a single interface. The HDMI specification defines three different regions for this purpose: video period, control period, and data island period (see Figure 4.10).

The video data period is in the active region when the actual pixels are sent. The data island period is when additional information like audio data is sent. The control period is when control data is sent. Guard bands and preambles are used to separate these regions. Table 4.9 shows the encoding type used and the data transmitted.

The video data is encoded using the standard TMDS encoding technique like DVI, where 8 bits of data is encoded as 10 bits. The data sent during the data island period is encoded using the TERC4 coding technique, where 4 bits of data are converted to 10 bits. Control codes sent during the control period are 10-bit encoded from 2 bits. And there are unique 10-bit codes for guard band. The actual data is either 2 or 4 or 8 bits, which is encoded as 10 bits, depending on whether it is control, Data Island, or video period, respectively.

The various periods are shown in detail in Figure 4.11. As per HDMI, the blanking period is classified as data island period or control period. The data island period is one during which data like audio is sent. The control period is the one where no data is sent, just the control information like HSync and VSync. The HSync and VSync occupy the 2LSB bits of the channel 0 during the entire blanking period, that is, during both the control and data island periods. They are encoded using control encoding during the control period and encoded as TERC4 encoding during the data island period along with the other data bits.

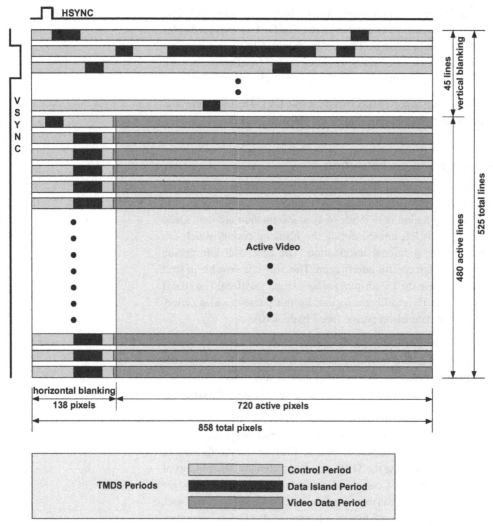

HSYNC

V S Y N C

45 lines vertical blanking

525 total lines

480 active lines

Active Video

horizontal blanking

138 pixels

720 active pixels

858 total pixels

TMDS Periods

Control Period

Data Island Period

Video Data Period

■ **FIGURE 4.10** Period placement in an HDMI frame.

Preambles for eight TMDS clocks followed by leading guard bands for two TMDS clocks are sent before both the data island period and video periods. The data island periods end with a trailing guard band as well. The codes for preamble and guard band for video and data island periods are different. The data island period contains at least one packet that is 32 TMDS clocks. The packet has header and packet data. The header is mapped to D2 of Channel 0. Hence the header contains a total of 32 bits, which are sent over 32 TMDS clocks. The data is mapped to both Channels 1 and 2, 4 bits

Table 4.9 Period Description

Period	Data Transmitted	Encoding Type
Video data	Video pixels	Video data coding (8 bits converted to 10 bits)
	(Guard band)	(Fixed 10-bit pattern)
Data island	Packet data - Audio samples - Infoframes HSync, VSync	TERC4 coding (4 bits converted to 10 bits)
	(Guard band)	(Fixed 10-bit pattern)
Control	Control - Preamble - HSync, VSync	Control period coding (2 bits converted to 10 bits)

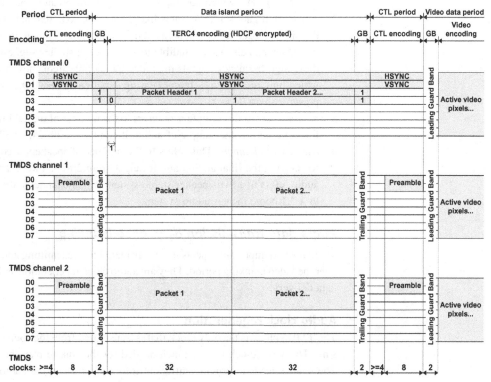

■ **FIGURE 4.11** TMDS periods and encoding.

each. So the total numbers of data bits are $(4 \times 32)+(4 \times 32)=256$. The video data will be sent during the active region and the number of TMDS clocks for which it will be sent will depend on the video mode being transmitted.

Audio basics

To understand how audio data is transmitted during the video blanking period it is important to understand the basic principles of audio. The digital audio stream is characterized by the following parameters:

- *Sampling rate*. Audio sampling rate refers to the rate at which the analog audio signal is sampled and converted to digital information. The same rate will be used when this digital audio stream is converted to analog audio during playback. As per the Nyquist theorem, the minimum sampling frequency required to capture a signal is twice the maximum frequency of the signal being captured. The audible frequency range is 20 Hz to 20 kHz for the human ear. Hence 40 kHz will be the ideal sampling rate for audio. The typical sampling rates of audio are 32, 48, or 44.1 kHz, which approximately satisfy the Nyquist theorem. There are higher sampling rates like 96 and 192 kHz that help capture the higher frequency components of the audio signal. Though these high frequency signals are not audible to human ears, studies suggest that these high frequency signals modulate the audible frequency signals, which may result in a different hearing experience, hence the need to preserve them.
- *Sample size*. Audio sample size refers to the number of digital bits used for representing one sample of audio. It can be 16, 20, 24, or 32 bits.
- *Number of channels*. This refers to the number of speakers used for playing the final audio data. It ranges from 2 to 8. Two-channel audio refers to a two-speaker stereo system. Eight channel audio refers to a 7.1 home theater audio system.

Audio data transmission over video blanking

There are two important aspects to be considered for transmitting audio data over the video blanking period. They are audio clock regeneration and audio data delivery.

Audio clock regeneration

The TMDS clock is the only clock that is transmitted from the source to the sink. The audio clock is very much needed for the sink to play the audio. The source needs to provide sufficient information about the audio clock to the sink so that the sink can regenerate the audio clock. The source can indicate to the sink the relationship of the audio clock with respect to the TMDS clock. With this information the sink regenerates the audio clock from the TMDS clock. The relationship between audio sampling rate (audio clock) and the TMDS clock is represented using the following equation as per the HDMI specification:

$$N/CTS = (128 \times F_{Audio})/F_{TMDS}$$

where N is any integer and CTS stands for cycle time stamp. Since the audio clock is in the kHz range whereas the TMDS clock is in the MHz range, the audio clock is multiplied by 128 to make the audio clock frequency comparable with that of TMDS clock. Otherwise the values of N and CTS will differ by a factor of 1000. If both the audio and TMDS clocks are coherent in the source, that is, both are derived from the same PLL within the source, then the N and CTS can be fixed values, and the specification recommends values for different audio and TMDS frequencies.

However, if the audio and TMDS clocks are noncoherent, that is, both are derived from two different PLLs in the source, then N will be fixed to the specification-recommended value and the CTS value is computed from the equation stated earlier. The same equation can be rearranged as follows:

$$N \times (1/(128 \times F_{Audio})) = CTS \times (1/F_{TMDS})$$

or

$$N \times (1/F_{Fast\_Audio}) = CTS \times (1/F_{TMDS})$$

where $F_{Fast\_Audio} = 128 \times F_{Audio}$, and F_{Fast_Audio} is a fast audio clock whose frequency is 128 times the audio clock. Since inverse of frequency $(1/F)$ indicates the time period of the clock, the above equation can be rewritten as shown below.

$$N \times (\text{Time Period of Fast\_Audio Clock}) = CTS \times (\text{Time Period of TMDS Clock})$$

From this it can be inferred that N number of fast audio clock periods is equal to CTS TMDS clock periods. So CTS can be computed by counting the number of TMDS clocks in every N Fast_Audio clocks as shown in Figure 4.12.

With this method any slight change in the audio clock frequency in the source will result in a change in the computed CTS value. This in turn allows the sink to closely track the slight changes in frequency of the audio clock at the source; this is very important for the audio playback at sink. The CTS value is computed by the source once every N Fast_Audio clocks and sent to the sink. The N and CTS are transmitted as an audio clock regeneration data island period and sent to the sink.

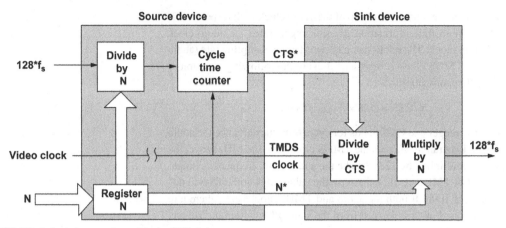

■ **FIGURE 4.12** Audio clock regeneration at sink using TMDS clock.

Audio data delivery rate

As described earlier, the audio data is transmitted during the data island periods, which occur during the blanking interval. So the source device has to buffer the incoming audio data till the blanking interval and transmit them. Since the horizontal blanking period occurs once every line, the source can potentially use the horizontal blanking period as an opportunity to transmit the buffered audio data. How many samples get buffered and will there be enough time in the blanking period to transmit the buffered samples? To address the first question we need to consider the rate at which the horizontal blanking period occurs and the audio sampling rate. The rate at which the horizontal blanking period occurs is nothing but line rate. For cases where the video line rate is greater than the audio sampling rate, then on an average one audio sample may be sent once every x number of video lines, where $x =$ (line rate)/(audio rate). For cases where the video line rate is less than the audio sampling rate, then on an average y number of samples may be sent once every video line, where $y =$ (audio rate)/(line rate). Table 4.10 suggests the required audio data delivery rate for 32- and 192-kHz audio stream for two different video modes.

Table 4.10 Audio Sample Delivery Rate Example

Video Mode	Line Rate (kHz)	Requirement for 32 kHz Audio	Requirement for 192 kHz
1920 × 1080@60	67.5	1 sample every 67.5/32 = 2.1 lines	192/67.5 = 2.88 samples every line
640 × 480@60	31.5	1 sample every 31.5/32 = 1 line	192/31.5 = 6 samples every line

Once the required audio data delivery rate is known it is now important to determine whether there is enough time or bandwidth in the horizontal blanking period of a given video mode to accommodate the same; in other words, it needs to be established whether there are enough TMDS clocks during the horizontal blanking period to transmit the required number of audio samples. The audio samples are transmitted as data island packets (DIPs). The transmission of one DIP takes 32 TMDS clocks as described earlier. As per the HDMI specification, a maximum of four audio samples can be sent using a single DIP for a two-channel audio stream, and for an audio stream with more than two channels, only one sample can be sent using a single DIP.

Hence, the maximum number of audio samples that can be sent during the horizontal blanking period:

=HBLANK width in terms of TMDS clocks/8 for an audio stream with two channels
=HBLANK width in terms of TMDS clocks/32 for an audio stream with more than two channels

This illustration assumes that only audio sample packets are sent during the horizontal blanking period, which may not be true in a real case. There might be other packets that also need to be sent during the horizontal blanking period. Also it does not account for preambles and guard bands that are required for DIP transmission. All these should be accounted for while doing this analysis. Basic audio like two-channel audio at sampling rates of 32, 44.1, and 48 kHz can be supported in almost all video modes. Audio with greater than two channels and rates up to 192 kHz are supported on selected video modes, depending on the bandwidth available during the blanking period. Special audio modes like IEC 61937 compressed audio (surround sound audio) are supported as well on specific video modes. A similar analysis as described earlier needs to be done to find out which audio modes can be supported for any given video mode.

Infoframes

Infoframes are defined as the miscellaneous information that is sent during the data island period. Auxiliary video (AVI) infoframes and audio infoframes are two of the common infoframes. The AVI infoframe contains additional information about the video mode being transmitted, like the pixel format and so forth. An audio infoframe contains additional information about the audio stream being sent, like the sampling rate, sample size, etc. The details of the infoframes are defined in the CEA-861D specification.

HDMI connector

The HDMI specification defines five different types of connectors: Type A, B, C, D, and E. Type A is the standard connector with 19 signals. Table 4.11 shows the pin assignment for a Type A connector.

Type B is larger than Type A and it has dual channels for supporting higher resolutions. It has a total of 29 signals. The pin assignments for the Type B connector are shown in Table 4.12.

Types C, D, and E have the same set of 19 signals as Type A, but the form factor and the pin assignments are different. Type C is smaller than Type A and is meant for mobile devices. Type D is smaller than Type C, and Type E is intended for automotive applications.

DisplayPort (DP)

DP is a relatively new standard developed by the VESA with the intention of having one unified and scalable specification for both internal and external displays. DP does not use a dedicated clock channel; there are only data channels (referred as lanes in the DP specification) with embedded clocking. The specification allows the number of lanes to be configurable, which helps save on cost and power, especially for internal displays. DP is also the first display interface to use packet-based data transmission for both video and audio data, which makes it possible to transmit either of them in the absence of the other (Figure 4.13).

Table 4.11 Pin Assignments for a Type A Connector.

PIN	Signal Assignment	PIN	Signal Assignment
1	TMDS Data2+	2	TMDS Data2 Shield
3	TMDS Data2−	4	TMDS Data1+
5	TMDS Data1 Shield	6	TMDS Data1−
7	TMDS Data0+	8	TMDS Data0 Shield
9	TMDS Data0−	10	TMDS Clock+
11	TMDS Clock Shield	12	TMDS Clock−
13	CEC	14	Utility
15	SCL	16	SDA
17	DDC/CEC Ground	18	+5 V power
19	Hot Plug Detect		

Table 4.12 Pin Assignments for a Type B Connector.

PIN	Signal Assignment	PIN	Signal Assignment
1	TMDS Data2+	2	TMDS Data2 Shield
3	TMDS Data2−	4	TMDS Data1+
5	TMDS Data1 Shield	6	TMDS Data1−
7	TMDS Data0+	8	TMDS Data0 Shield
9	TMDS Data0−	10	TMDS Clock+
11	TMDS Clock Shield	12	TMDS Clock−
13	TMDS Data5+	14	TMDS Data5 Shield
15	TMDS Data5−	16	TMDS Data4+
17	TMDS Data4 Shield	18	TMDS Data4−
19	TMDS Data3+	20	TMDS Data3 Shield
21	TMDS Data3−	22	CEC
23	Reserved (N.C. on device)	24	Reserved (N.C. on device)
25	SCL	26	SDA
27	DDC/CEC Ground	28	+5 V Power
29	Hot Plug Detect		

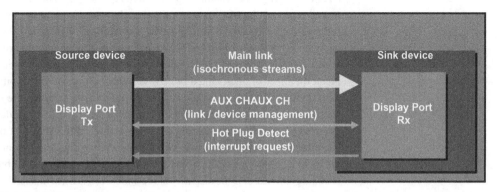

■ **FIGURE 4.13** DisplayPort data transport channels.

The DP interface consists of

- *Main link*. This is used for transmitting the video or/and the audio data.
- *AUX channel*. This is a sideband interface used for serving the DDC purpose along with link management.
- *Hot Plug Detect*. This is used for both Plug and Unplug Event detection at the source as well as being used as a means to notify the sink using the interrupt mechanism.

Main link operation

The main link comprises a maximum of four data lanes, which use differential signaling and ANSI 8b/10b encoding; there is no separate clock lane. The sink recovers the clock from the data lanes itself. The ANSI 8b/10b encoding technique is designed in such a way that the following rules are met:

1. The maximum number of consecutive 1s or 0s is 5.
2. The average DC value of the lane is 0. This is done by restricting the difference between the number of 0s and 1s in any 10-bit symbol to ± 1.

This verifies that sufficient transitions are in the line for the receiver to able to recover the clock from the data lanes and also continually align itself with the incoming stream. There is no specific algorithm defined for this technique; it is more of a lookup table implementation. The appropriate 10-bit (symbol/code) is picked up from a lookup table based on the incoming 8-bit data. The 10-bit code for every 8-bit data is represented as $D_{x.y}$, where x is the lower 5 bits of the 8-bit data and y is the upper 3 bits of the 8-bit data.

The 5-bit is encoded as 6-bit code and the 3-bit is encoded as 4-bit code. If the incoming 8 bits are named A through H starting from LSB to MSB, then the lower 5 bits ABCDE are encoded as *abcdei* and the upper 3 bits FGH are encoded as *fghj*. The resulting 10-bit code is *abcdeifghj*. The LSB is transmitted first and the MSB is transmitted in the end. Figure 4.14 shows the same as to how the 8-bit data (character) is converted to 10-bit data (symbol) and vice versa. Each 10-bit code can have two possible representations, one with a greater number of zeros, called *negative disparity* and the other with a greater number of ones, called *positive disparity*. Some of the 10-bit codes have an equal number of 1s and 0s. The encoding engine within the source keeps track of the difference in ones and zeros from the 10-bit codes sent previously; this is called running disparity. Based on this information, the encoder picks up the 10-bit code with the appropriate disparity for encoding the incoming 8-bit data. The rules for picking up the appropriate 10-bit code are straightforward. If the disparity of the code word is zero, that is, the number of ones and zeros are equal, then there is no choice and the running disparity remains unchanged. If the previous running disparity is positive, then the code with negative disparity is chosen and vice versa.

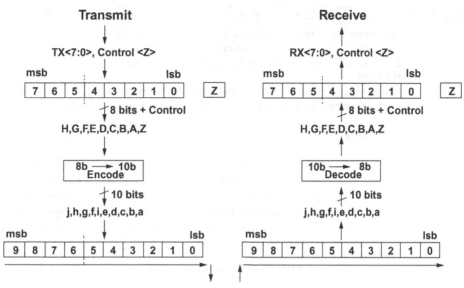

■ **FIGURE 4.14** Character to symbol mapping.

Apart from data symbols, there are unique 10-bit codes called *control* or *comma* or *K codes* (or *characters*) represented as $K_{x.y}$. These 10-bit codes do not match with any of the 10-bit data codes; that is, the control codes use a different set of lookup tables than the data. These codes are used for fields like delimiters, guard bands, and blanking start (BS) of a protocol. These codes are used by the sink to recover the clock from the incoming stream.

AUX channel operation

AUX channel is a bidirectional interface using Manchester II encoding with embedded clocking operating at a bit rate of 1 Mbps. In Manchester II encoding, 0 is represented by 0 to 1 transition in the middle of the bit period and 1 is represented by 1 to 0 transition in the middle of the bit period as shown in Figure 4.15.

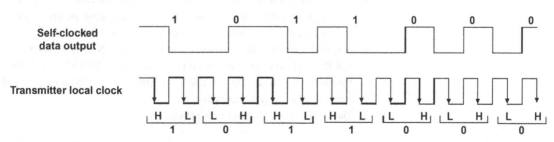

■ **FIGURE 4.15** Self-clocking with Manchester II coding.

Pre-charge	Sync pattern (SP)	End of SP	Transaction	STOP

Pre-charge – 10 to 16 consecutive zeros in Manchester II code.

Sync pattern – 16 consecutive zeros in Manchester II code.

End of SP – High for 2 μs followed by low for 2 μs (this is not as per Manchester-II).

Transaction – Request or response transaction in Manchester II code.

STOP – High for 2 μs followed by low for 2 μs (this is not as per Manchester-II).

■ **FIGURE 4.16** AUX channel transaction format.

Source is the master as it initiates all transactions by transmitting a request transaction. After completing the request transaction, the source releases the buses for the sink to respond back with a response transaction. On receiving a response transaction, the source can start another request transaction. The source waits for a maximum timeout period of 300 μs for the sink to respond back.

All AUX transactions have the format shown in Figure 4.16.

The transaction consists of different fields and the fields are different for request and reply transaction. The request transaction from source device consists of the following fields:

1. Command is a 4-bit field using which the source indicates to the sink the type of transaction like read, write, or I^2C over AUX, etc.
2. Address is 20-bit field indicating the address from where the source either wants to write or read from.
3. Length field is 1 byte in size, and it indicates the number of bytes of data the source wants to read or write from the specified address. The maximum value of length is 16 bytes. If more than 16 bytes of data needs to be written or read then the source needs to initiate multiple AUX channel transactions.
4. Data field is valid only for write transactions and the number of bytes of data sent for write should match with what is specified in length field.

The reply transaction from sink consists of the following fields:

1. Command is a 4-bit field using which the sink indicates to the source what type of actions the sink has taken in response to the last request transaction from the source. Some of the actions can be acknowledgment (ACK) or no acknowledgment (NACK or defer (DFR). And there are different types of responses for I^2C over AUX transactions as well. The 4-bit reply command is padded with 4 bits of zeros to align to byte boundary.
2. Data field exists for read transactions only.

The AUX channel is used for doing either a native AUX channel transaction or an I²C over AUX transaction. The former is used for link maintenance and so forth, which involves reading and writing registers from the sink. The Display Port Specification defines registers for various purposes like link maintenance, which are called DPCD (display port configuration data). The latter is used for EDID enumeration.

Link training

As described earlier, the DP does not have a dedicated clock lane, and the sink has to recover the clock from the data lanes and align to the 10-bit symbol boundary. So before the source can start the transmission of the video/audio data, it needs to make sure that the sink is in the state wherein it can extract the 10-bit symbols accurately. The process by which the source and sink make sure that the main link is ready for actual data transmission is called *link training*. This is a two-step process that includes clock recovery, symbol boundary, and inter-lane alignment. The DP specification stipulates a predefined set of symbols called the *training pattern* for each of these steps. The different training patterns are

1. Training pattern 1—this pattern consists of sending D10.2 characters continuously without scrambling. The sink uses this pattern for clock recovery.
2. Training pattern 2—this pattern consists of the sequence of the characters sent over and over continuously: K28.5 −, D11.6, K.28.5+, D11.6, D10.2, D10.2, D10.2, D10.2, D10.2, D10.2, without scrambling. The sink uses this pattern for channel equalization, alignment to symbol boundary, and inter-lane alignment.

The source configures the main link to transmit the appropriate training pattern in a repetitive manner. There are DPCD registers in the sink that can be configured using the AUX channel to indicate which training pattern is being sent on the main link. Also, there are DPCD registers that indicate whether the sink is able to successfully recover the clock or align to the symbol boundary. In addition to this, the sink can request for a change in the electrical parameters of the main link. The source can read all this information and take appropriate action and request the sink to retry. A maximum of five retries is allowed for each step. If the sink is able to successfully recover the clock and align to symbol boundary, then the link training is successful and the link is ready for video or audio transmission.

The electrical parameters that a sink can request for altering during link training include voltage swing and pre-emphasis. The voltage swing refers to the maximum peak to peak voltage between the *p* and *n* (V_{DIFF}) of the differential lines of each data lane as indicated in Figure 4.17. The various swings are 400, 600, 800, and 1200 mV.

The pre-emphasis indicates the reduction in the peak to peak voltage in the case of consecutive 1s or 0s for the subsequent bits with respect to the first bit as indicated in Figure 4.18.

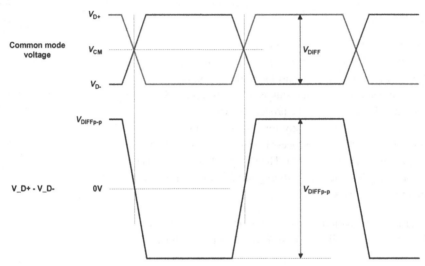

■ **FIGURE 4.17** Sample waveform of voltage swing.

Pre-emphasis = 20·Log($V_{DIFF-PRE}/V_{DIFF}$)

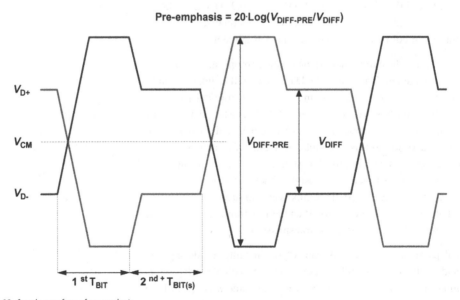

■ **FIGURE 4.18** Sample waveform of pre-emphasis.

Figure 4.18 indicates that the first bit shows the actual Full Voltage Swing ($V_{\text{DIFF-PRE}}$) whereas for the second bit onwards the peak to peak voltage is reduced (V_{DIFF}) slightly below the maximum voltage swing. This is helpful in making the transitions from one polarity to another faster. This is represented in dB and the supported values are 0, 3.5, 6, and 9.5 dB.

Link operation

The main link is operated at a fixed frequency called the *link frequency* or *link rate* irrespective of the video mode being transmitted. There are two supported link rates of 270 and 162 MHz, which are referred as high bit rate (HBR) and reduced bit rate (RBR) respectively. Since 10 bits of data are sent for every link clock, the data rates are 2.7 and 1.62 Gbps. Since the sink has only the link clock, it needs to regenerate the pixel clock from the link clock. The same is true for audio clock or sampling rate. In order to communicate the relationship of the video/audio clock with respect to the link clock, fields called M and N are used. Both audio and video have their own M and N parameters, which are 24 bits in size:

$$M_{\text{Video}}/N_{\text{Video}} = P_s/L_s \text{ for Video}$$
$$M_{\text{Audio}}/N_{\text{Audio}} = 512 F_s/L_s \text{ for Audio}$$

where P_s is the pixel clock, L_s is the link clock, and F_s is the audio sampling rate.

If the pixel clock and the link clock are coherent, then M_{Video} and N_{Video} are constant values such that they satisfy the above equation. If the pixel clock and the link clock are not coherent, then N_{Video} is fixed to a constant value of 0x8000 and M_{Video} is computed. Rearranging the above equation yields the following:

$$M_{\text{Video}} \times (1/P_s) = N_{\text{Video}} \times (1/L_s)$$
$$M_{\text{Video}} \times (\text{Time Period of Pixel Clock}) = N_{\text{Video}} \times (\text{Time Period of Link Clock})$$

From the above equation it can be concluded that M_{Video} can be computed by counting the number of pixel clock cycles in every N_{Video} link clock cycle. The same method is applicable for the audio clock too. The source sends video M and N values in a packet called main stream attributes (MSA), and the audio M and N values are sent using an audio time stamp packet. The sink regenerates the pixel and audio clocks using the appropriate M and N values.

The DP specification does not mandate that all the four lanes be used. It can be operated in 1-lane (x1), 2-lane (x2), or 4-lane (x4) modes. The maximum possible bandwidth that can be supported in each of these lane configurations is listed in Table 4.13.

Table 4.13 Maximum Possible Bandwidth in Different Lane Configurations

	X1 (Mbytes/s)	X2 (Mbytes/s)	X4 (Mbytes/s)
RBR	162	324	648
HBR	270	540	1080

The following equation can be used to determine whether a given video mode will be supported in a particular lane configuration:

$$\text{Pixel Clock} \times \text{Bytes Per Pixel} \leq \text{Link Clock} \times \text{Number of Lanes}$$

Micro packets—Transfer units

DP uses the concept of packet-based data transmission, like Ethernet. The two main types of packets are the transfer unit (TU) and secondary stream. The TU (see Figure 4.19) is used for the transmission of the video data and is considered as a micro packet. The size of a TU can vary from 32 to 64 link clocks.

The number of link clocks for which pixels are sent in a TU is given by the following equation:

$$((\text{Pixel Clock} \times \text{Bytes Per Pixel})/(\text{Link Clock} \times \text{Number of Lanes})) \times \text{TU Size}$$

For example, if the above equation gave a value of 32 for a TU size of 64, then for 32 link clocks of a TU, actual valid pixels are sent and the remaining are filled with certain stuffing symbols. And if the above equation gave a fraction value of 32.25, then once in every four TUs valid pixels are sent for 33 link clocks and the remaining three TUs are packed with valid pixels for 32 link

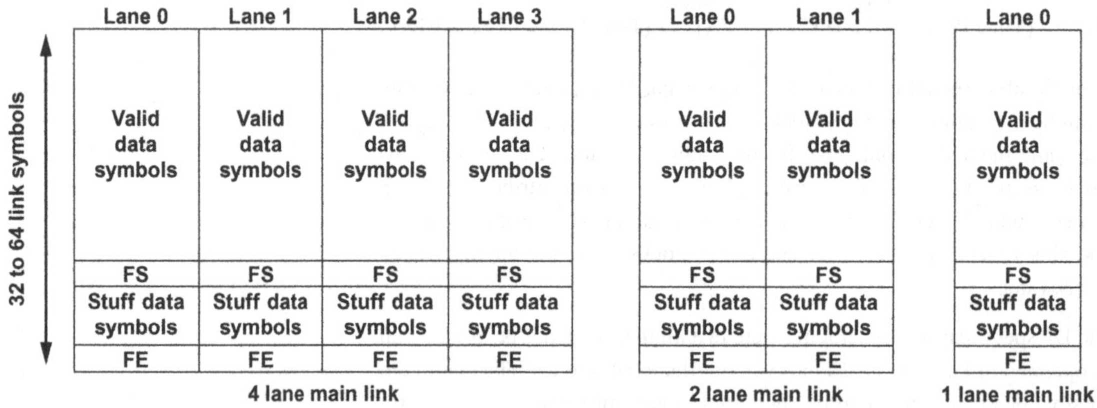

■ **FIGURE 4.19** Transfer unit.

clocks. The organization of TUs for different lane configurations is shown in Figure 4.19. The starting and ending of stuffing is marked by special symbols called Fill Start and Fill End, shown as FS and FE in the figure.

Figure 4.20 depicts the transmission of the TUs. The BE (blanking end) marks the beginning of the active video period, and BS marks the beginning of the blanking period. Between these two there are many TUs that are transmitted that may be partially filled with pixels as per the video mode and lane configurations. The dark-shaded region indicates the pixels within the TU and the light-shaded region indicates the stuffing codes within the TU. The packetization of pixels within the TUs will depend on the bits per pixel and the number of lanes. Figure 4.21 shows the way in which pixels are

■ **FIGURE 4.20** The transmission of TUs.

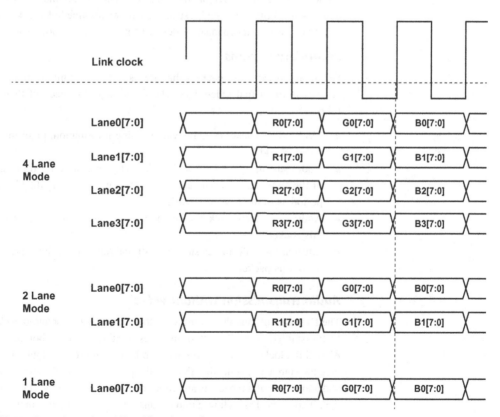

■ **FIGURE 4.21** Timing diagram for 24 BPP in different lane configurations.

packed in a TU for 24 BPP in different lane configurations and the 8 bits of data before being encoded as 10 bits. The link clock is shown for completeness but it is not actually sent to the sink. The number of pixels sent varies as per the number of lanes. In the case of 4-lane mode, 4 pixels are sent for every 3 link clocks. And the same pattern repeats for the subsequent clocks. Similarly in the case of 2-lane mode and 1-lane mode, 2 pixels and 1 pixel are sent respectively for every 3 link clocks.

Figure 4.22 shows the way in which pixels are packed in a TU for 18 BPP (6 bpc) in different lane configurations. The diagram shows the 8 bits of data before being encoded as 10 bits. The link clock is shown for completeness but it is not actually sent to the sink. Since the number of bits per color (6 in this case) is less than the actual data bits that can be sent on a lane on each link clock (8), the actual useful data straddles the subsequent clock periods. The same can be seen in Figure 4.22. The number of pixels sent varies according to the number of lanes. In the case of 4-lane mode, 16 pixels are sent for every 9 link clocks. And the same pattern repeats for the subsequent clocks. Similarly, in the case of 2-lane mode and 1-lane mode, 8 pixels and 4 pixels are sent respectively for every nine link clocks. A similar pixel packing mechanism is defined for other pixel formats as well.

Secondary streams

Secondary streams are packets that are used for transmitting information other than the actual video data. The functionality of each of them is as follows:

- MSA—This packet contains video timing information, pixel format, video M and N.
- Audio stream packet—This secondary stream is used for transmission of the audio packets. The audio packets are sent during the blanking period as in the case of HDMI.
- Audio time stamp packet—Audio M and N values are transmitted using this packet.
- Inoframes—Infoframes such as AVI and audio infoframes are sent using this packet.

Audio transmission without video

The DP transmits an idle pattern when there are no video or audio packets to be transmitted. The idle pattern involves sending a BS symbol once every 8192 link clocks. If only audio needs to be transmitted without any video, then the source transmits the BS as in idle pattern, and audio stream packets and audio timestamp packets are inserted as required after the BS symbol in the idle pattern. This is how audio is transmitted without video in the case of

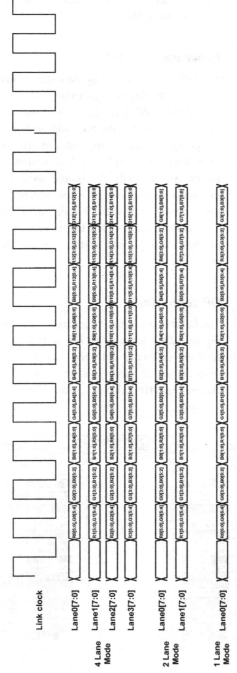

■ FIGURE 4.22 Timing diagram for 18 BPP in different lane configurations.

DP. One may ask what the usage of such a feature is. If one wants to use one's TV/monitor just for listening to music sometimes, then it can be achieved using DP.

Other uses of HPD

The HPD signal is used by the source to detect a plug and unplug of a DP monitor. On plug a high value is observed on the HPD line, and on unplug the HPD line has a low value. The source monitors for these values to determine the plug and unplug events. Apart from this, during the plugged condition the sink can drive a low value on the HPD for a short time period and then drive a high value again. The sink uses this mechanism to notify the source about important events. For example, the source can be notified of the loss of main link synchronization using this mechanism, and the source can reinitiate link training in response. The source detects such a low pulse with width less than 2 ms as a notification interrupt, and anything greater than 2 ms is seen as an actual unplug event.

DP connector

The DP connector pin assignments in the source and sink side are shown in Figure 4.23. The main link signals that comprise four differential pairs get reversed within the cable; hence the sink side pin assignments are different from the source with respect to these signals. All other signals, like AUX channel and HPD, have the same signal assignment in both the source and sink.

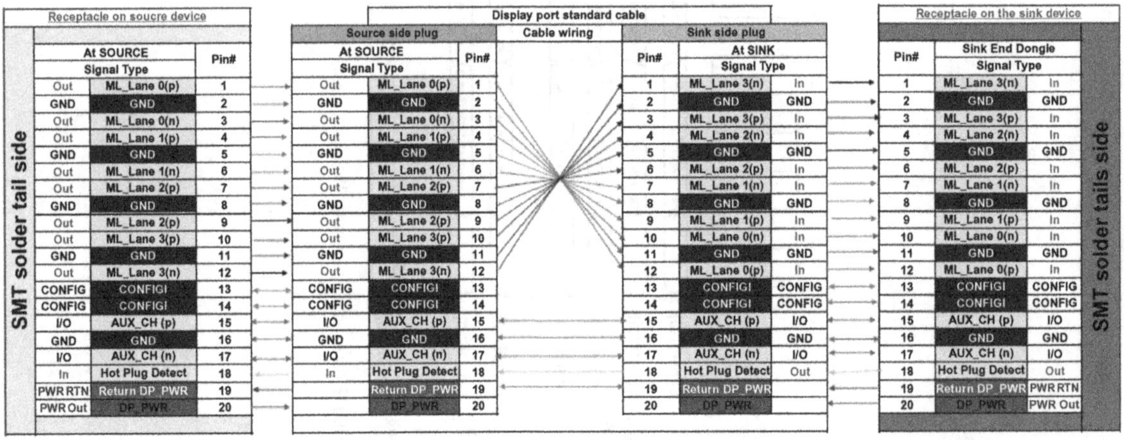

Receptacle on source device (SMT solder tail side)			Source side plug (At SOURCE)			Cable wiring	Sink side plug (At SINK)			Receptacle on the sink device (SMT solder tails side)		
Dir	**Signal Type**	**Pin#**	**Dir**	**Signal Type**	**Pin#**		**Pin#**	**Signal Type**	**Dir**	**Pin#**	**Signal Type**	**Dir**
Out	ML_Lane 0(p)	1	Out	ML_Lane 0(p)	1		1	ML_Lane 3(n)	In	1	ML_Lane 3(n)	In
GND	GND	2	GND	GND	2		2	GND	GND	2	GND	GND
Out	ML_Lane 0(n)	3	Out	ML_Lane 0(n)	3		3	ML_Lane 3(p)	In	3	ML_Lane 3(p)	In
Out	ML_Lane 1(p)	4	Out	ML_Lane 1(p)	4		4	ML_Lane 2(n)	In	4	ML_Lane 2(n)	In
GND	GND	5	GND	GND	5		5	GND	GND	5	GND	GND
Out	ML_Lane 1(n)	6	Out	ML_Lane 1(n)	6		6	ML_Lane 2(p)	In	6	ML_Lane 2(p)	In
Out	ML_Lane 2(p)	7	Out	ML_Lane 2(p)	7		7	ML_Lane 1(n)	In	7	ML_Lane 1(n)	In
GND	GND	8	GND	GND	8		8	GND	GND	8	GND	GND
Out	ML_Lane 2(p)	9	Out	ML_Lane 2(p)	9		9	ML_Lane 1(p)	In	9	ML_Lane 1(p)	In
Out	ML_Lane 3(p)	10	Out	ML_Lane 3(p)	10		10	ML_Lane 0(n)	In	10	ML_Lane 0(n)	In
GND	GND	11	GND	GND	11		11	GND	GND	11	GND	GND
Out	ML_Lane 3(n)	12	Out	ML_Lane 3(n)	12		12	ML_Lane 0(p)	In	12	ML_Lane 0(p)	In
CONFIG	CONFIGI	13	CONFIG	CONFIGI	13		13	CONFIGI	CONFIG	13	CONFIGI	CONFIG
CONFIG	CONFIGI	14	CONFIG	CONFIGI	14		14	CONFIGI	CONFIG	14	CONFIGI	CONFIG
I/O	AUX_CH (p)	15	I/O	AUX_CH (p)	15		15	AUX_CH (p)	I/O	15	AUX_CH (p)	I/O
GND	GND	16	GND	GND	16		16	GND	GND	16	GND	GND
I/O	AUX_CH (n)	17	I/O	AUX_CH (n)	17		17	AUX_CH (n)	I/O	17	AUX_CH (n)	I/O
In	Hot Plug Detect	18	In	Hot Plug Detect	18		18	Hot Plug Detect	Out	18	Hot Plug Detect	Out
PWR RTN	Return DP_PWR	19		Return DP_PWR	19		19	Return DP_PWR		19	Return DP_PWR	PWR RTN
PWR Out	DP_PWR	20		DP_PWR	20		20	DP_PWR		20	DP_PWR	PWR Out

■ **FIGURE 4.23** DP connector pin assignments.

DP defined a scalable display interface for both the internal and external displays. Source and sink devices can actually choose the number of lanes and the link rates based on the maximum video and audio mode that they want to support. This provides the source/sink vendors with a lot of flexibility while designing the hardware. Also, using the same protocol for both the internal and external display helps reuse the same DP IP.

INTERNAL DISPLAY INTERFACE

The displays that are always attached to devices like laptops, Ultrabook devices, tablets, or smartphones are called *internal display*. They cannot be plugged into or unplugged from the device using a cable like the external displays. In the case of external displays, the display interface is used only for conveying the timing and pixel information, and it does not deliver the actual power required to drive the display. The external display is usually connected to an external power source. However, this is not possible in the case of internal displays. Because the internal displays are attached to the devices like phones and tablets, they are powered by the batteries in these devices. This is one of the key differentiating factors between the internal and external displays. Hence different internal display interface standards are defined to address the challenges that arise due to the above-stated requirement.

Overview

The two main challenges any internal display interface has to tackle are internal display power consumption and internal display power sequencing. Since the internal displays are powered by the battery of the device, they directly impact battery life of the device—that is, how long the device can operate without recharging the battery. Hence it is important that any internal display interface is efficient in terms of power and has good power-saving techniques defined as part of the specification. The internal display power sequencing (also known as panel power sequencing, since most of the internal displays are LCD panels) arises due to the rules involved in powering up and powering down a panel. Some of the internal display specifications address this as well. Before we dive into the various internal display interface specifications, let us understand these two challenges and a few possible generic solutions as well.

LCD operation

Most of the internal displays are liquid crystal displays, or LCDs. Let us begin by understanding how an LCD panel works. The LCD panels have a backlight that produces white light. This white light, when passed through the liquid crystal, can produce different colors based on the orientation of the crystals. And the orientation of the crystals can be controlled by subjecting the crystals to different electric fields, which in turn can be controlled by varying the voltage applied across them.

From this it is clear that in order to drive an internal LCD, the following elements are required:

1. Display interface—This is the interface with which the pixel and timing information is transmitted to the LCD.
2. Panel power—This is the power required for converting the digital pixel information received through the display interface to the appropriate voltage across the liquid crystal to produce the desired color.
3. Backlight power—This is the power required to drive the backlight of the LCD.

Panel power sequence

LCD panels have stringent rules for the order in which the above three elements should be enabled or disabled during powering up or powering down the panel, with appropriate delays between each step. Otherwise it can damage the LCD panels. This is known as *panel power sequencing*. The most commonly used sequence for powering up the panel is to turn the panel power on, send the pixels through the display interface, and then turn the backlight power on. The reverse order is followed while powering down the display. There are delays between each of these steps that need to be met as part of the sequence. Typically all external displays have a separate microcontroller to take care of these, whereas in the case of internal display adding a separate microcontroller for this purpose is not a viable option as this adds to the cost. Hence this is taken care of by the display controller of the SoC of the actual device (smartphone, laptop, tablet, and so on). This helps save on cost. The display controller can easily control when to send pixels on the display interface. And to control the panel and backlight power, separate enable signals can be used. These enable signals can act as enables for power delivery circuitries.

Panel power saving

The next challenge for internal displays is *power-saving techniques*. This is quite an important aspect of internal display because the internal display is one of the main sources for draining the device battery. Various power-saving techniques can be used for internal displays that help increase the battery life. Some of these techniques are explained in the next section.

Backlight power reduction

The backlight power can be reduced by using a special signal called the pulse width modulation (PWM) signal. This signal is used after the panel is powered up. The PWM signal indicates to turn the backlight on when it is high and to turn it off when it is 0. Driving a 1 will result in full brightness of the backlight and 0 will turn the backlight off completely. The user can see the image only when the backlight is on. By switching between on and off at a fast rate, the end user perceives a change in the brightness of the image being viewed and not any other visual artifacts. Hence by controlling the rate of switching, the brightness can be controlled and this helps save the power consumed by the backlight of the panel. To compensate for the change in the brightness of the backlight, the brightness in the image being displayed itself can be modified such that the end user does not perceive any noticeable change due to the change in the brightness of the backlight. Special algorithms are developed that take into account the brightness level in the image being displayed and the ambient lighting conditions under which the user is viewing the image to determine the optimum backlight intensity and the image brightness required such that there is no perceivable difference for the user while viewing the image under a given condition. This technique is referred to as the Intel Display Power Saving Technology (Intel DPST) and it helps save backlight power to a great extent.

Remote frame buffer

All display interfaces continuously fetch the frame buffer data (the buffer in the system memory that contains the image to be displayed) and transmit the same to the display, as shown in Figure 4.24.

Even when the image being displayed doesn't change that frequently (static image), like when the user is reading something, the process of fetching the frame buffer from system memory and transmitting the same through the

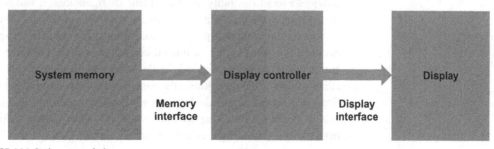

■ **FIGURE 4.24** Display memory fetches.

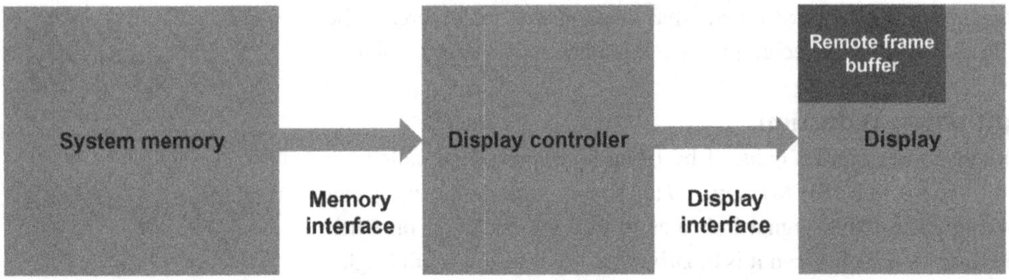

■ **FIGURE 4.25** Remote frame buffer within display.

display interface has to continue at the refresh rate. This causes unnecessary power consumption by system memory, memory controller, display controller, and interface. If it is possible to store the frame buffer itself within the display, then we can shut down the system memory, memory controller, display controller, and interface when the image being displayed is static, and the display can fetch from its buffer and display the image. Such a buffer within the display is referred to as a *remote frame buffer* (RFB) (see Figure 4.25).

The SoC can do selective updates to this buffer if there are only a few regions in the image that need an update or it can send an entire new frame whenever there is a major change. This is implemented as command mode in the MIPI display interface and as panel self refresh (PSR) in the eDP display interface. These will be discussed in detail in the later sections.

So far we have discussed the basic fundamentals of the internal displays. There are three different types of internal display interfaces: LVDS, eDP, and MIPI DSI. The LVDS evolved as an internal display interface for laptops and notebooks. This is still the most widely used internal display interface for these types of devices. It does not support the RFB mechanism described earlier. MIPI DSI evolved as an internal display interface for smartphones. It supports the RFB mechanism and it is very power efficient. The eDP is an internal display interface based on the DP protocol, and it provides a lot of flexibility in terms of lane configurations; it also supports the RFB mechanism, and it can be used for all types of devices. The connectors for the internal displays are usually not talked about in the specification. Some specifications define the connectors with the pin assignments but do not define the mechanical details. Some specifications do not define either of them. And many times the panel vendors deviate from the specification in order to make sleek panels with smaller connectors and shorter cables. Hence the following sections, which talk about these interfaces in detail, will not discuss the connectors of the internal display.

Low Voltage Differential Signaling (LVDS)

LVDS has been used as a display interface for internal display for a very long time for notebooks, netbooks, and tablets. This interface is comprised of pixel clock channel and data channels. In one pixel clock period, 7 bits of data are transmitted serially on the data channels. No encoding technique like TMDS or ANSI 8b/10b is used for the data; instead, actual data is transmitted. The data that is transmitted on the data channels includes the following:

- Pixel data—Pixel data in either 24 or 18 BPP format is used.
- HS—The horizontal sync bit is high during the horizontal blanking period.
- VS—The vertical sync bit is high during the vertical blanking period.
- DE—The DE bit is high during the video active period.

The maximum frequency of the LVDS clock in the LVDS interface is 112 MHz, which results in a maximum bitrate of 784 Mbps on the data channels.

The interface can be operated in either 18 BPP or 24 BPP pixel formats, also referred to as 6 bits per color mode or 8 bits per color mode, respectively. The channel data mapping for these modes is shown in Figure 4.26. The interface can be operated in dual channel modes with more data channels. In dual channel mode, the even and odd pixel data are sent on different data channels as shown in Figure 4.27.

The 18 BPP mode helps reduce the number of data channels compared to the 24 BPP mode. The reduction in the number of data channels helps save the cost and power by reducing the cable and connector size. However, the pixel format used by the SoC while creating the frame buffer in memory is 24 BPP most of the time. Reducing the number of bits per color component from 8 to 6 is done by dropping the LSB 2 bits; this introduces a quantization error, which can result in visual artifacts. A special algorithm known as dithering is used to reduce such artifacts. Dithering is an algorithm that diffuses such quantization errors to the neighboring pixels, thereby reducing the visual artifacts and making the image look as much as possible like the original image.

LVDS panel power sequencing

The power sequencing for the LVDS panels requires proper sequencing of three parameters: power supply, LVDS interface, and backlight enable. They are turned on with power supply first, then LVDS interface, and backlight in the end; when powering down the order is reversed.

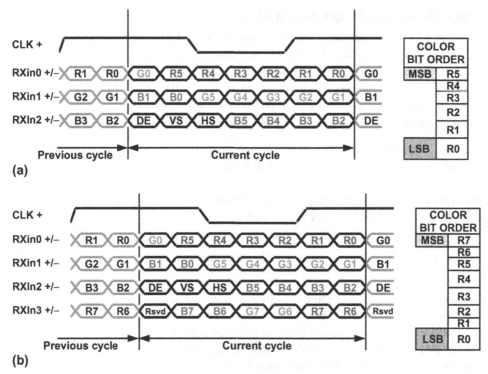

■ **FIGURE 4.26** (a) LVDS 6-bit and (b) 8-bit single channel data mapping diagrams.

There are specific delay requirements while powering up and powering down as shown in Figure 4.28. The various parameters are explained here:

- T1 is the time for the power supply to reach 90% of its rated value from 10% of its rated value during the powerup sequence.
- T2 is the delay between the power supply becoming stable and the LVDS interface getting enabled during powerup sequence. A minimum value of 0 indicates that the LVDS interface can be enabled immmediately after the power supply is stable.
- T3 is the delay between disabling the LVDS interface and turning off the power supply during the powerdown sequence. A minimum value of 0 indicates that the power supply can be turned off immmediately after disabling the LVDS interface.
- T4 is the minimum delay required before the powerup sequence can be initiated, once again after powering down the panel.
- T5 is the delay between the LVDS interface getting enabled and turning on the backlight during the powerup sequence.
- T6 is the delay between turning off the backlight and disabling the LVDS interface during the powerdown sequence.

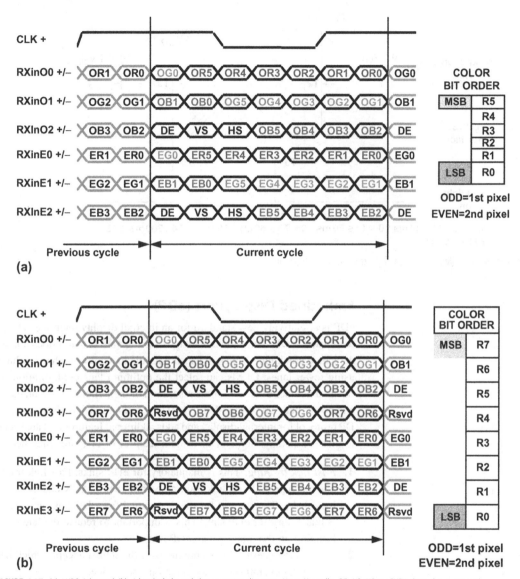

■ **FIGURE 4.27** (a) LVDS 8-bit and (b) 6-bit dual channel data mapping diagrams. Note: Normally, DE, VS, HS on EVEN channel are not used.

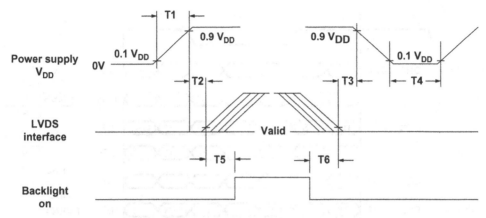

$0.5\,ms \leq T1 \leq 10\,ms : 0 \leq T2 \leq 50\,ms : 0 \leq T3 \leq 50\,ms : 500\,ms \leq T4 : 200\,ms \leq T5 : 200\,ms \leq T6$

■ **FIGURE 4.28** Timing diagram: LVDS panel power up and down sequencing.

Embedded DisplayPort (eDP)

eDP is a relatively new standard for an internal display interface. The protocol used is DP. The main advantage of using eDP is that it does not use a dedicated clock lane, which makes it possible to achieve higher resolutions using fewer lanes. OEMs can select the optimum lane configuration (link rate and number of lanes) for the resolution of their choice based on the cost and power budgets permissible. The eDP specification defines the use of the DP protocol for internal display and adds additional features that are specific to internal display. Some of the salient features of eDP are listed here:

1. *Fast link training.* This is mandatory for eDP sinks. Fast link training is a mechanism to do link training without using AUX channel handshake. The support for this is optional for a DP sink, whereas it is mandatory for eDP sink. This is important to reduce the latency of bringing up the image on an eDP panel.
2. *Panel self refresh.* This technique uses the RFB concept, which helps reduce the overall power consumed in the device.

In this method the frame buffer controller in the source looks for opportunities to save power whenever a static image is being displayed and informs the eDPTX about the same, as shown in Figure 4.29. The eDP TX uses a special secondary stream data packet called PSR SDP to communicate to the sink to store the incoming frame in the RFB and henceforth display the image from the same. This will allow the source to save power by powering down the memory eDP main link and so forth. Whenever the frame buffer controller

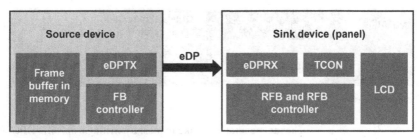

■ **FIGURE 4.29** Panel self refresh system level diagram.

detects that the image has changed, it requests the eDP TX to either send the entire new frame or to do selective updates in the RFB if the image has changed partially. The frame buffer controller detects the writes done on the frame buffer and initiates the eDP TX to take necessary action. TCON is the timing controller used by all panels. It is shown in Figure 4.29 for completeness.

PSR entry is signaled by using a special PSR Secondary Data Packet (SDP) (see Figure 4.30). Secondary streams are used for transmitting various information like audio data and infoframe data as already explained in "DisplayPort" section. Secondary streams are transmitted during the blanking period. Similarly the PSR SDP is a secondary stream data packet defined for PSR, and it follows all the packetization rules that other secondary stream packets use.

The PSR SDP packet has to be sent before the PSR setup time. This setup time is in terms of the number of scanlines before the first active scanline. On receiving this SDP, the sink stores the entire frame following this SDP. After transmitting the first eight blanking lines in the next frame, the main link of eDP can be shut down, and the sink will switch to RFB to show the image on the display. The PSR exit is shown in Figure 4.31. The AUX

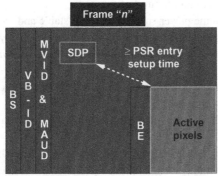
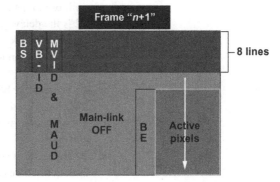

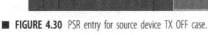

■ **FIGURE 4.30** PSR entry for source device TX OFF case.

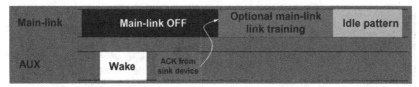

■ **FIGURE 4.31** PSR exit link management.

channel generates a wake request by writing a specific DPCD register in the sink. The sink responds back with an ACK, following which the source can train the main link if required and transmit idle patterns. During this period the sink will continue to display the image from the RFB, and it will exit PSR only when it receives a PSR SDP requesting the sink to exit PSR as part of a video frame.

eDP panel power sequencing

The eDP panel power sequencing is similar to LVDS panel power sequencing, the main difference being the addition of link training as part of the sequence and the addition of the HPD signal. The HPD signal is driven by the sink at the appropriate time, unlike the other signals that are driven by the source. The HPD signal is not used for plug or unplug detection, since an eDP panel will always be attached to the device. As described earlier, the sink can use this to notify the source in case of events like loss of sync. Apart from this the sink is also responsible for automatic black video during link training.

The normal eDP interface power up/down sequence is shown in Figure 4.32. The various parameters are explained here:

- T1 is the time for the power supply to reach 90% of its rated value from 10% of its rated value during the powerup sequence. Or in other words, it is the rise time of the power supply, and it has a minimum limit of 0.5 ms and a maximum limit of 10 ms.
- T2 is the delay between the power supply becoming stable and the automatic black video generation by the sink. This action is done by the sink. The minimum value of this delay can be 0 and the maximum value is 200 ms.
- T3 is the delay between the power supply becoming stable and the HPD signal assertion by the sink. This action is done by the sink. The minimum value of this delay can be 0 and the maximum value is 200 ms.
- T4 is the delay between the HPD signal assertion to the link training initiation. There is no minimum or maximum requirement for this delay. It is dependent on the source device.

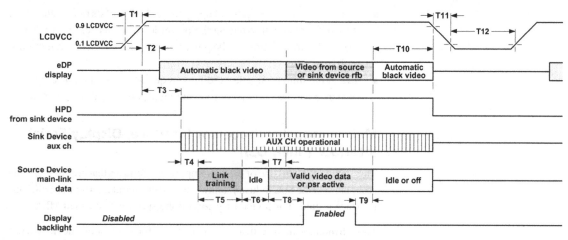

■ **FIGURE 4.32** eDP interface power up/down sequence, normal system operation.

- T5 is the actual duration of the link training. There is no minimum or maximum requirement for this delay.
- T6 is the actual duration for which an idle pattern is sent by the source after successful link training. There is no minimum or maximum requirement for this delay.
- T7 is the delay between the valid video data from the source to the actual video on display. The minimum value of this delay can be 0 and the maximum value is 50 ms.
- T8 is the delay between the valid video data from the source and the assertion of backlight enable. The backlight can be turned on only when there is an active video stream. So during link training and idle pattern transmission, the backlight enable is not asserted. There is no minimum or maximum requirement for this delay. But the source needs to make sure there is valid video data when the backlight is enabled.
- T9 is the delay between the de-assertion of backlight enable and the end of valid video data on the main link. There is no minimum or maximum requirement for this delay. However, the source needs to make sure there is valid video data as long as the backlight is enabled.
- T10 is the delay between the end of valid video on the main link to the powering down of the panel. During this time, the sink is supposed to maintain automatic black video on the display. The minimum value of this delay can be 0 and the maximum value is 500 ms.

- T11 is the time for the power supply to reach 10% of its rated value from 90% of its rated value during the powerdown sequence. Or in other words, it is the fall time of the power supply and it has a maximum limit of 10 ms.
- T12 is the minimum delay (500 ms) required before the powerup sequence can be initiated once again after powering down the panel.

Mobile Industry Processor Interface, Display Serial Interface (MIPI, DSI)

This section explains one of the very important interfaces that are gaining wide acceptance in the SoC industry, known as Display Serial Interface or DSI. This interface is guided by an industry consortium specification called MIPI.

Though initially the specification was a guiding force for mobile devices, it has also become very popular among other industries that rely on low-power SoC designs. The DSI is commonly targeted at LCD, ranging from 3 to 12 in. in size. This interface specifies the connections, connectors, and the protocol between the sources of image data (generally called the host) to the display device. The DSI specification is very closely coupled with the physical layer I/O specification called DPHY (from MIPI) but is not mandated. DPHY is the physical layer used for point-to-point serial communication at very high speed (approximately 1.5 GHz). Though DPHY is a serial physical layer, DSI is not strictly a serial communication mechanism, but rather it defines a quasi-serial protocol where multiple DPHY can be utilized to achieve required throughput greater than a single DPHY. The DSI specification builds on existing specifications by adopting pixel formats and a command set defined in the display command set (DCS) specification from MIPI.

About MIPI

MIPI (Mobile Industry Peripheral Interface) is an organization that defines interface specification for mobile devices. There are different types of interfaces in a mobile device like the display interface, camera interface, Bluetooth interface, storage, high speed, low latency interface, etc. The MIPI consortium defines the specifications for all of these types of interfaces in such a manner that it is consistent and reusable.

Adaptation of a set of standard hardware interfaces will produce peripheral hardware products from multiple vendors that work seamlessly with numerous processor and system-on-a-chip products. Since software is an integral component of many of these interfaces, some level of software standardization from MIPI enhances this improved interconnectivity. A common set of interfaces maximizes design reuse, drives innovation, and reduces time-to-market.

MIPI pervades many of the interfaces used inside mobility devices like tablets and mobiles and covers peripherals like mass storages, displays, cameras, audio devices, microphones, debug interfaces, off chip storages, modems, power management chips, and so on.

Introduction

DSI specifies the interface between a host processor and a peripheral display device. An attempt is being made here to give the essence of the DSI specification. The essential terminology used inside the standard are used, so as to make the reader understand the basic concept. Since DPHY is the essential component on which the DSI specification is structured, an introduction to essential concepts of DPHY is also discussed. Readers are requested to refer to the specification of DPHY and DSI for more detailed description.

DSI caters to the needs of two basic display architectures. In one, it is assumed that the peripheral device (display here) has no buffer for either few lines of pixels or for the entire frame; that is, the display displays the pixels as and when they arrive. This is called video mode display. In another display architecture, the system has a frame buffer, where the final display occurs when all the data for the complete frame is available with the peripheral's display. This is called *command mode display*. The two types of display are shown in Figure 4.33.

In video mode display, the host provides pixel data and timing information to the display device, and pixels are displayed as and when they are received, whereas in command mode display, the data for the whole frame is provided to the display, which buffers the data and then data from the buffer is displayed to the screen, based on the timing provided by the timing controller known as TCON. Here essentially the data is transferred in bursts between the host and the display device, and TCON relieves the host from timing management. Since command mode display requires a buffer to store the incoming data, it is costly, but since the data transfer is in bursts, the physical layer can be put in low power mode to save energy.

Both command mode and video mode display must be compatible with the commands in the MIPI specification of DCS; however, vendor-specific commands are also possible.

Advantages for display with DSI features

Tradeoff is a way of life. The tradeoff between cost and performance is fundamental to any system design. DSI provided various advantages required for the displays in the mobile industry. A serialized interface gives it a low pin count (a very big motivation for the SoC industry) in comparison to the

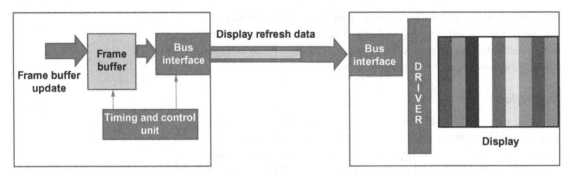

Video mode

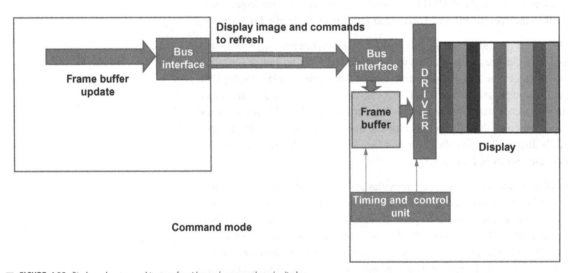

Command mode

■ **FIGURE 4.33** Display subsystem architecture for video and command mode display.

prevalent parallel interfaces. The low power coupled with low pin count is a very big advantage where battery life is one of the key differentiating factors in the competitive mobile space. DSI also supports a wide range of architectures for displays of video and command mode type and also a quasi-video-command type where the buffer can be very small. The rich sets of commands of DCS cater to most of the needed functionality. Bidirectionality of the lanes also provides various debug capabilities and manufacturer-specific functionality and configurability. Some of the other key advantages are

■ Supports both command and video mode architectures
■ Maintains all functionality of legacy parallel interfaces

- Low-power signaling support and bidirectionality
- Standard display formats supported
- Lane-scalable
- Low operational power and very low standby power
- Bidirectional data capability, to support command mode
- Up to 12-cm conductor length, with connectors and flex cable
- Excellent EMI rejection and low emissions, low error rate
- Minimizes pin count and cost—no exotic circuit design
- Protocol supports multiple displays

System design

The basic DSI system

Figure 4.34 shows a basic DSI system. It consists of a differential clock lane and multiple differential data lanes. The data lanes can be a minimum of one to multiple (*N*). As per the protocol specified for DSI, the first data lane has to be bidirectional. All the signaling over these lanes is differential except for Lane 0, which can have single-ended signaling as well.

From a conceptual viewpoint, a DSI-compliant interface sends pixels or commands to the peripheral and can read back status or pixel information from the peripheral.

DSI functional layers

A DSI organizes the interface into several functional layers as shown in Figure 4.35.

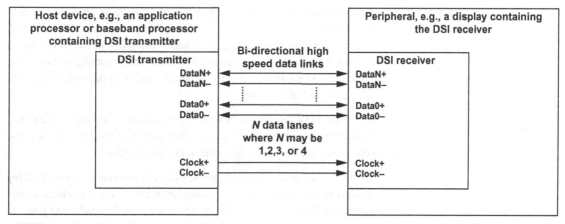

■ **FIGURE 4.34** A basic DSI subsystem.

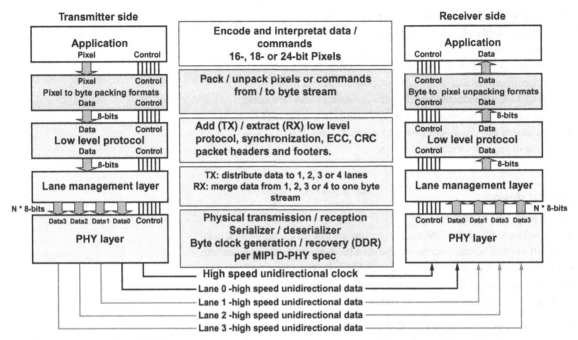

■ **FIGURE 4.35** Functional layers of DSI subsystem.

The lowest layer of the DSI subsystem is the *physical layer* or PHY. PHY is responsible for physical transmission and reception of data bits serially. The transmission can occur in HS (high speed) or LP (low power or low speed) mode. The DPHY specification of PHY mandates that reverse transmission must always occur in LP mode. The PHY layer also transmits the clock in a dedicated clock lane. The clock recovery and data recovery are the part of PHY layer. Data that can be split into multiple lanes is transmitted to the sink in sync with the clock, and the relationship is defined in the DPHY specification, which we will discuss in some detail in the coming sections. The PHY layer is also responsible for serializing and de-serializing the parallel data presented to each lane.

Above the PHY layer is the *lane management layer*. This layer is responsible for merging all the data that are coming from different lanes into a single contiguous and serialized data stream in correct order.

The *LLP layer* or the *low-level protocol layer* is responsible for packetizing the data received from the higher layer and presents a well-packetized payload for the lane management layer. This layer adds the header and footer to the payload and adds ECC and CRC bytes to make a complete packet.

The topmost layer is the *application layer*. This layer receives the pixels from the graphic processor or any other image-forming layers. This layer correctly interprets and buffers the pixel data. This layer also provides the host with an interface to provide commands to the display device. The pixel or command data from this layer is suitably packed or unpacked into a byte stream by the layer below it.

Physical basis for high-performance DSI

The DPHY specification provides the physical basis for the high performance of the DSI subsystem. The HS mode (high-speed mode) of data transfer is used by DPHY for only the forward direction of data transfer. The differential data lane uses both the edges of the clock to sample the data. Thus essentially the data rate is twice the clock rate. The clock rate can vary anywhere from 100 MHz to 1 GHz, putting a throughput of 200 Mbps to 2 Gbps of data. The number of data lanes required is dependent on the data throughput required for the display, which depends on the display resolution and pixel resolution, that is, on the size of the display and the "bits per pixel" of the display.

There is another mode of DPHY that is called LP mode. This mode is used mostly for command transmission in the forward direction or any data read by the host from the display in the reverse direction. The voltage swing and the data recovery for LP mode is quite different from the HS mode, which will be discussed in "DPHY signaling for HS and LP mode" section.

Figure 4.36 shows the various phases of an HS data transmission in relation to the clock. It must be noted that all the phases are not mandatory.

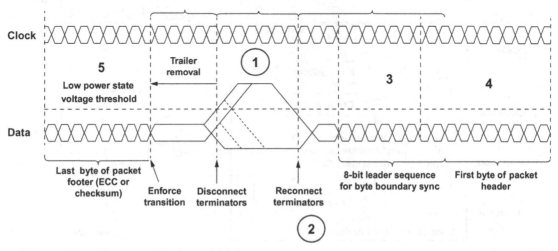

■ **FIGURE 4.36** Phases of data transmission in relation to the clock.

As can be seen from Figure 4.36, after the LP stage (1), which could be idle or the LP data transmission stage, the matched termination for HS (50 Ω) are enabled (2), and there is an 8-bit sequence of synchronization (3) before transmission of the first HS (4) packet data. Similarly, at the end of packet transmission (5), generally with ECC or checksum, there is a toggling of the last byte of data, and a cooling period called EoT (end of transmission sequence) is allowed.

More details about the DPHY internal structure follows, which allows the type of transmission just described.

DPHY signaling for HS and LP mode

The typical sub-block of a DPHY IP shows (Figure 4.37) blocks for LP for both the transmit and receive directions. Similarly, there is an HS sub-block. There is a block (Rt) for termination load. Also there is a CD (contention

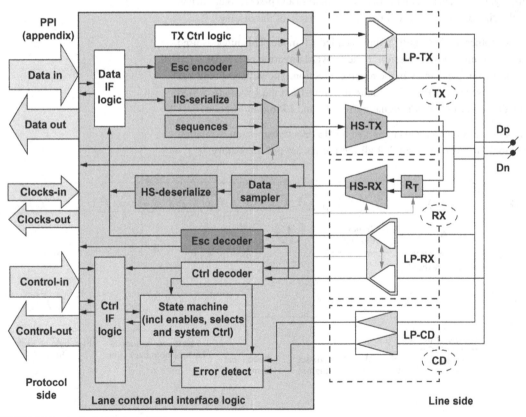

■ **FIGURE 4.37** Functional sub-blocks in the DPHY.

detection) block. The LP and HS are mutually exclusive. The CD and Rt blocks are active once the DPHY is active. HS blocks mostly have the Serializer and Deserializer, which convert the Parallel PPI (PHY peripheral interface) into serial and vice versa. Similarly, on the LP side the main section is the Esc encoder (to be discussed in a subsequent section), which takes care of the LP serialization and bus turnaround (BTA). The rest of the control block state machine is responsible for power management and contention detection.

The DPHY on the master and slave are interconnected through differential traces of physical connection. The interconnect between DPHY and PHY adaptation layer of DSI block is standardized and called PPI (PHY peripheral interface).

The termination values are selected based on the mode of data transmission. For HS a termination of 50 Ω and for LP mode 100 Ω is selected. For the sake of understanding further, let us discuss the voltage levels and the definition of various states and the nomenclature thereof.

Figure 4.38 shows the voltage level for both HS and LP states of transmission. As can be seen in Figure 4.38, the LP signal swings from 0 to 1.1 V (1). The voltage minimum and maximum are also shown. For HS state signaling, the common mode voltage is 250 mV with 140 mV of voltage swing (2). Various voltage margin available to the system is also shown. Kindly note the voltage swing range for the HS common mode voltage and the corresponding voltage swing possible.

Based on the state of individual lane pairs' voltage level, DSI defines various lane states for HS and LP as shown in Figure 4.39.

Based on this description of states, let's see how the different transmission of data occurs in different modes. Thus we will discuss LP transmission mode, the HS data transmission mode. All different phases of each mode of transmission will also be discussed.

LP transmission mode

LP data transmissions, which is generally used for commands from the hosts and response from the display, are also called escape mode commands. Low-power data transmission is comprised of three phases:

- "escape mode entry phase" (LP11 → LP00 → LP01 → LP00)
- the "Entry command or data phase"
- and "stop state" (LP10 → LP11)

which are shown in Figure 4.40 in different *zones*.

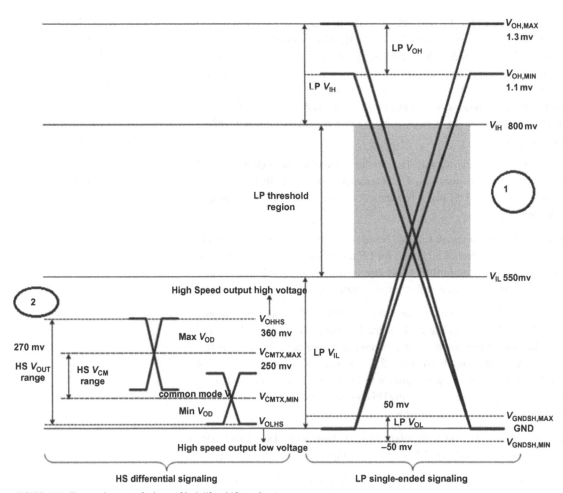

■ **FIGURE 4.38** Showing the range of voltages of both HS and LP signaling.

State code	Line voltage levels		High-speed	Low-power	
	Dp-line	Dn-line	Burst mode	Control mode	Escape mode
HS-0	HS low	HS high	Differential-0	N/A, Note 1	N/A, Note 1
HS-1	HS high	HS low	Differential-1	N/A, Note 1	N/A, Note 1
LP-00	LP low	LP low	N/A	Bridge	Space
LP-01	LP low	LP high	N/A	HS-Rqst	Mark-0
LP-10	LP high	LP low	N/A	LP-Rqst	Mark-1
LP-11	LP high	LP high	N/A	Stop	N/A, Note 2

■ **FIGURE 4.39** Lane states and their code.

LP-11>10>00>01>00>01>00>10>00>...

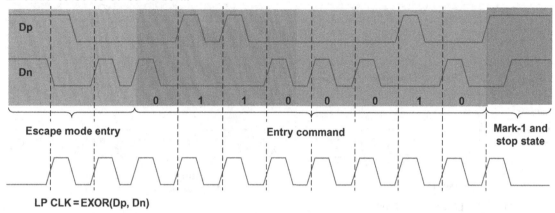

LP CLK = EXOR(Dp, Dn)

■ **FIGURE 4.40** LP escape mode signaling with embedded clock.

In the figure it is also shown how the clock is extracted by performing an XOR operation on the data lane DP and DN. The LP mode of data transmission is used for bidirectional data transmission of commands and as per the DSI protocol only lane 0 is allowed to go for LP data transmission. This is also called the *escape mode*.

Sometimes the escape mode command requires a reversal of direction of transmission from slave to master, so a method of BTA is mandated by DPHY. The sequence of BTA is shown in Figure 4.41, which goes through the sequence of state of LP11 → LP10 → LP00 → LP10—LP00, and during the drive overlap period in Figure 4.41, a change of drive direction occurs from master to slave. The various timing parameters shown in the figure are described in detail in the specification.

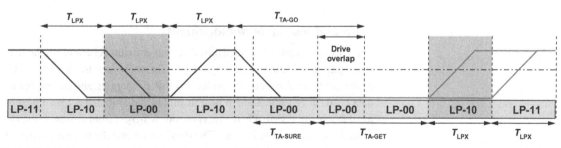

■ **FIGURE 4.41** Showing state transition during a bus turnaround (BTA).

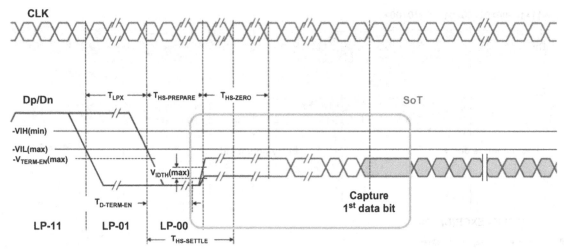

■ **FIGURE 4.42** Start of HS transmission (SoT).

Similarly, for high-speed (HS) mode there are different phases:

- SoT (start of transmission) phase,
- data transmission phase,
- and EoT.

as shown in Figures 4.42 and 4.43, respectively. Detailed descriptions can be found in the specification. As can be seen, that transition from idle LP11 state to HS mode requires some mandatory timing requirement. Kindly also note that during transition from LP to HS mode a change in termination occurs. HS mode starts with a sync pattern that enables the receiver's clock to synchronize with the receiving data This is called SoT period. Similarly, during EoT there is a resting time mandated by THS-Trail after which switchover to LP states is possible.

Data and multilane management

The previous section described how data transmission in LP and HS mode is made physically. But this description was for single-lane transmission. In a multilane DSI subsystem, the additional task of merging and demerging of data needs to be done. Data to be transmitted is distributed across the lanes available in the system. On the receiver side a merger from different lanes to a single stream of data is done. The distribution and the merger happen at byte level, as shown in Figure 4.44.

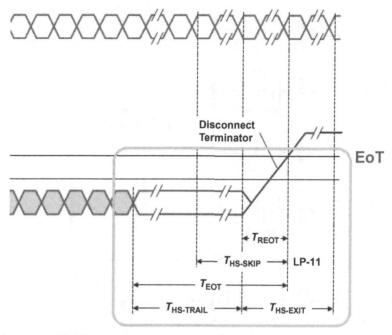

■ **FIGURE 4.43** End of transmission of HS (EoT).

Byte packaging in a multilane transmission

Till now we have discussed the mechanism of transmission of data that can be split either in a single lane or in multiple lanes in a serialized fashion. When a stream of data is split into multiple lanes, the organization of the split byte is important and predefined. The ordering of the bytes in multilane transmission happens as shown in Figure 4.45, which shows how the data bytes should be organized when the number of bytes to be transmitted is integral or odd multiple of the number of lanes available to the DSI.

Packet sequence of DSI

Once the data are split for multilane transmission or even in the case of single-lane transmission, data needs to be packetized for transmission, and there is some packet overhead that is added for proper delimitation and differentiation of data and the purpose of data. These overhead will be discussed when we discuss the packet structures. There are two types of packet defined in DSI, namely short packet (SP) and long packet (LP). A typical HS data transmission sequence consisting of LP and SP can consist of a mix of LP and HS signaling (see Figure 4.46). Kindly note that idle state when there is no data to transmit is an LP11 state.

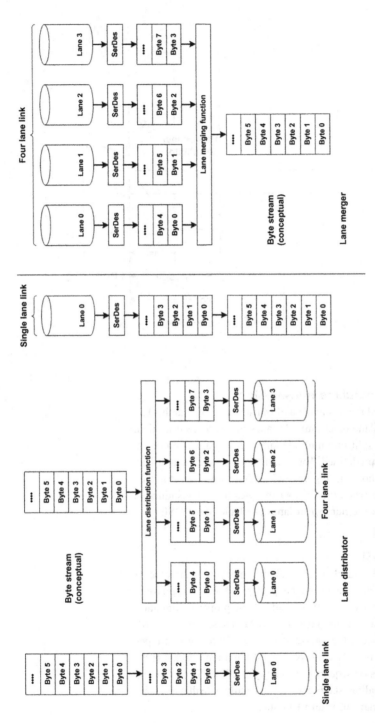

■ **FIGURE 4.44** Showing lane distribution and lane merging in relation with byte sequence.

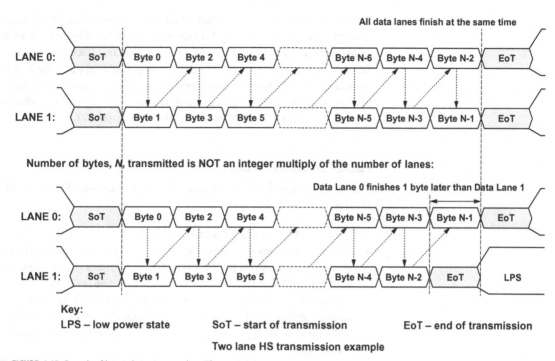

All data lanes finish at the same time

Number of bytes, *N*, transmitted is NOT an integer multiply of the number of lanes:

Data Lane 0 finishes 1 byte later than Data Lane 1

Key:

LPS – low power state   SoT – start of transmission   EoT – end of transmission

Two lane HS transmission example

■ **FIGURE 4.45** Example of byte ordering in a two-lane HS transmission.

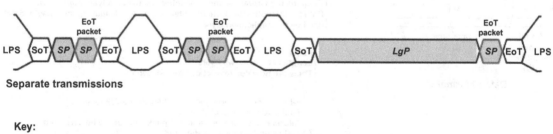

Separate transmissions

Key:

LPS – Low power state   SP – Short packet

SoT – Start of transmission   LgP – Long packet

EoT – End of transmission

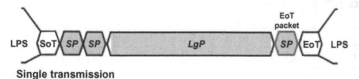

Single transmission

■ **FIGURE 4.46** A typical transmission with different packets.

Before describing the packet format, let us also discuss the concept of virtual channel so that the different fields of the packet can be understood. A virtual channel (VC) is used for a multidrop data destination attached to a single DSI interface. It must be noted that the VC is applicable not to a single lane but to a single DSI interface, which may include multiple lanes. A VC identifier is used to identify the data destination, and accordingly a DSI hub section routes the data. The DSI protocol allows up to four VCs.

Packet structure of DSI

The various fields of a DSI packet are listed and explained here (see also Figures 4.47 and 4.48):

- DATA ID byte = virtual channel ID + data type
- DATA TYPE conveys information on packet size, so the receiver can identify the end of one packet and beginning of the next. Two types of DATA TYPES are defined, one for forward direction (examples include display refresh pixel stream, commands, frame buffer update) and another for reverse direction (examples include READ responses, ACK, and error report). Again, a data packet can be the short or long data packet type.

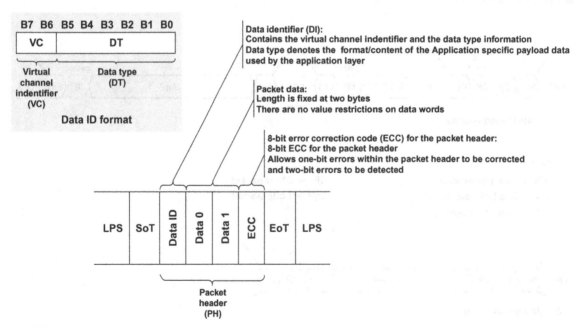

■ **FIGURE 4.47** Short data packet structure.

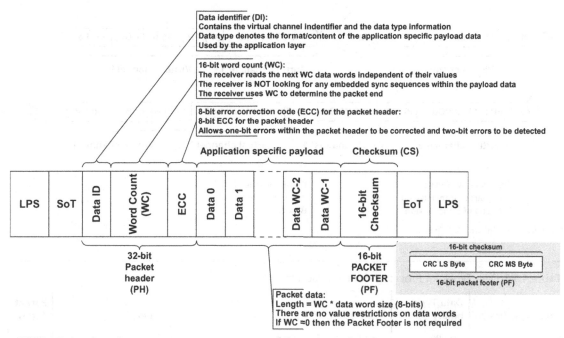

Data identifier (DI):
Contains the virtual channel indentifier and the data type information
Data type denotes the format/content of the application specific payload data
Used by the application layer

16-bit word count (WC):
The receiver reads the next WC data words independent of their values
The receiver is NOT looking for any embedded sync sequences within the payload data
The receiver uses WC to determine the packet end

8-bit error correction code (ECC) for the packet header:
8-bit ECC for the packet header
Allows one-bit errors within the packet header to be corrected and two-bit errors to be detected

Application specific payload   Checksum (CS)

| LPS | SoT | Data ID | Word Count (WC) | ECC | Data 0 | Data 1 | ⋯ | Data WC-2 | Data WC-1 | 16-bit Checksum | EoT | LPS |

32-bit Packet header (PH)

16-bit PACKET FOOTER (PF)

16-bit checksum

| CRC LS Byte | CRC MS Byte |

16-bit packet footer (PF)

Packet data:
Length = WC * data word size (8-bits)
There are no value restrictions on data words
If WC =0 then the Packet Footer is not required

■ **FIGURE 4.48** Long data packet structure.

- Packet header includes ECC, the error-correction code (optional in reverse direction).
- Long packets include checksum protection (optional in reverse direction).

In actual data transmission, various combinations of SP and LP occur back to back. In general, SP are used for configuration and command and timing transmission for display, and LP are used for pixel data. An example of interleaved transmission of LP and SP is shown in Figure 4.49.

The main purpose of DSI packets are to enable a DSI system to transmit pixel and timing information. These pixels can be transmitted in a long packet, and commands and timing are mostly transmitted in a short packet. The various packet types are identified with their code. The data type code for various type of packets is shown in Table 4.14.

Pixel data transmission
The transmission of pixel and timing data depends on the mode or architecture of the display device. In a simple command mode display, pixel data are encapsulated in a long packet of DSI. An example of two types of pixel data is shown in Figure 4.50.

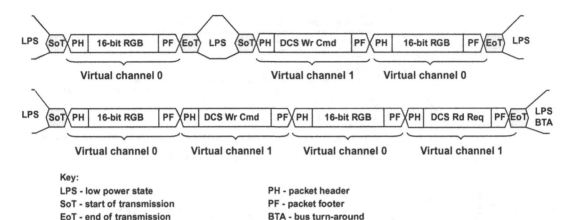

Key:

LPS - low power state   PH - packet header
SoT - start of transmission   PF - packet footer
EoT - end of transmission   BTA - bus turn-around

■ **FIGURE 4.49** Interleaved data streams.

Table 4.14 Data Types for Processor-Sourced Packets

Data Type, hex	Data Type, binary	Description	Legends	Packet Types
01h	00 0001	Sync Event, V Sync Start		Short
11h	01 0001	Sync Event, V Sync End		Short
21h	10 0001	Sync Event, H Sync Start		Short
31h	11 0001	Sync Event, H Sync End		Short
08h	00 0001	End of Transmission packet (EoTp)		Short
02h	00 0010	Color Mode (CM) Off Command		Short
12h	01 0010	Color Mode (CM) On Command		Short
22h	10 0010	Shut Down Peripheral Command		Short
32h	11 0010	Turn On Peripheral Command		Short
03h	00 0011	Generic Short WRITE, no parameters		Short
13h	01 0011	Generic Short WRITE, 1 parameter		Short
23h	10 0011	Generic Short WRITE, 2 parameters		Short
04h	00 0100	Generic READ, no parameters		Short
14h	01 0100	Generic READ, 1 parameter		Short

(Continued)

Table 4.14 Data Types for Processor-Sourced Packets *(Continued)*

Data Type, hex	Data Type, binary	Description	Legends	Packet Types
24h	10 0100	Generic READ, 2 parameters		Short
05h	00 0101	DCS Short WRITE, no parameters		Short
15h	01 0101	DCS Short WRITE, 1 parameter		Short
06h	00 0110	DCS READ, no parameters		Short
37h	11 0111	Set Maximum Return Packet Size		Short
09h	00 1001	Null Packet, no data		Long
19h	01 1001	Blanking Packet, no data		Long
29h	10 1001	Generic Long Write		Long
39h	11 1001	DCS Long Write/write_LUT Command Packet		Long
0Eh	00 1110	Packed Pixel Stream, 16-bit RGB, 5-6-5 Format		Long
1Eh	01 1110	Packed Pixel Stream, 18-bit RGB, 6-6-6 Format		Long
2Eh	10 1110	Loosely Packed Pixel Stream, 18-bit RGB, 6-6-6 Format		Long
3Eh	11 1110	Packed Pixel Stream, 24-bit RGB, 8-8-8 Format		Long
x0h and xFh, unspecified	xx 0000	DO NOT USE		
	xx 1111	All unspecified codes are reserved		

Once a display system is ready to receive various types of packets with timing and pixels information, a video display structure needs to be created. This may be dependent on the display manufacturer. The various regions of a display structure are shown, which are comprised of region line HSync (horizontal sync), HBP (horizontal back porch), HFP (horizontal front porch), etc. In video mode, various packets are transmitted that exactly coincide with the timing of the exact display (see Figure 4.51).

Thus a short or long packet is transmitted to replicate the traditional video timing adhering to the various display structure regions as shown in Figure 4.52.

Therefore various data types are defined for the packets, which are used to replace the analog position of the display structure in digital format as shown in Figure 4.53.

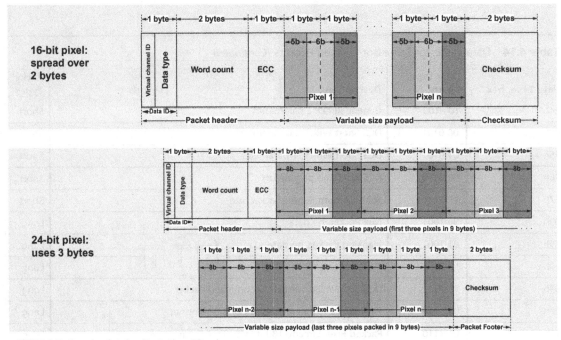

■ **FIGURE 4.50** Examples of pixel packing inside as DSI packet.

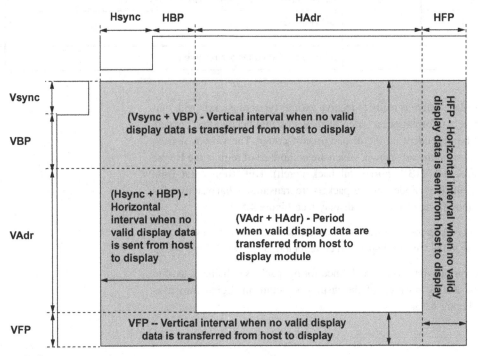

■ **FIGURE 4.51** Display timing parameters.

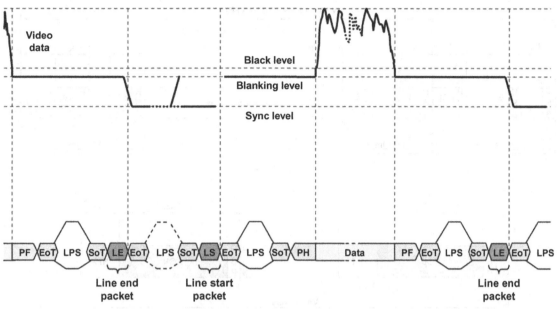

■ **FIGURE 4.52** Packet position with respect to traditional (analog) video timing.

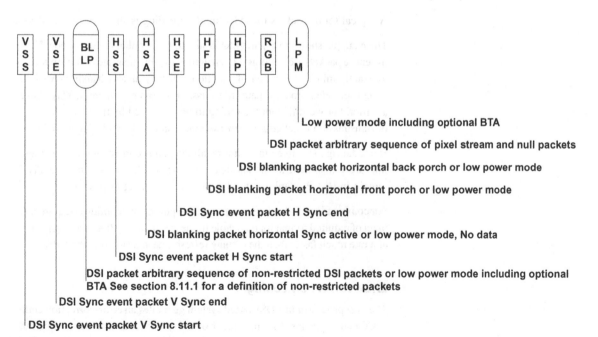

Low power mode including optional BTA

DSI packet arbitrary sequence of pixel stream and null packets

DSI blanking packet horizontal back porch or low power mode

DSI blanking packet horizontal front porch or low power mode

DSI Sync event packet H Sync end

DSI blanking packet horizontal Sync active or low power mode, No data

DSI Sync event packet H Sync start

DSI packet arbitrary sequence of non-restricted DSI packets or low power mode including optional BTA See section 8.11.1 for a definition of non-restricted packets

DSI Sync event packet V Sync end

DSI Sync event packet V Sync start

■ **FIGURE 4.53** Video mode interface timing legend.

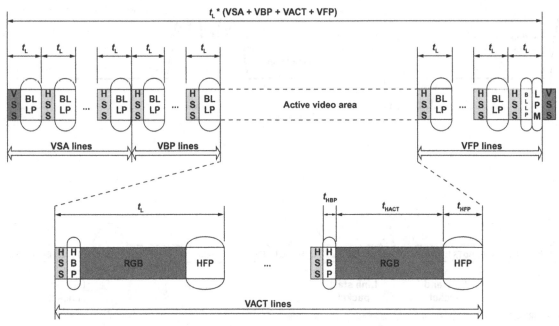

■ **FIGURE 4.54** A complete video mode with various DSI packets.

A typical set of packets for video mode signaling is shown in Figure 4.54.

The example shown in Figure 4.54 is said to be a nonburst video mode. Here the entire packets are so adjusted in length, based on their rate of transmission, to exactly mimic the timing of traditional display and its banking. A slight variant in video mode of data transmission is possible where displays have a buffer, but we still want the timing to be controlled by the host. This type of transmission is called *burst transmission* and is shown in Figure 4.55.

Here, during the active video timing, all the pixels corresponding to a single line are transmitted in the shortest possible duration, such that the PHY remain in BLLP (blanking line LP) mode for as long as possible.

For command mode displays, no stringent requirement of timing is required. A pixel of required format for the whole frame is sent line by line in a long packet at a rate much faster than the display refresh rate and stored in the buffer.

Error-correction coding

The host processor in a DSI-based system generates an error-correction code (ECC) and appends it to the header of every packet sent to the peripheral. The ECC takes the form of a single byte following the header bytes. The

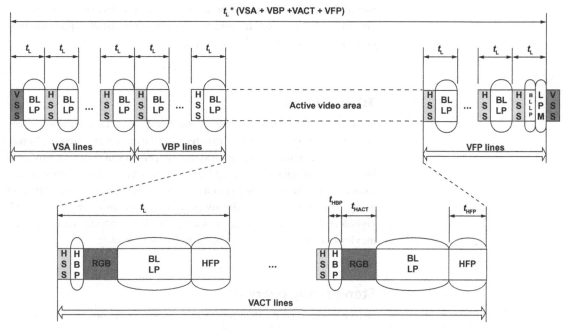

■ **FIGURE 4.55** Video mode interface timing: burst transmission.

ECC byte provides single-bit error correction and 2-bit error detection for the entire packet header. ECC is always generated and appended in the packet header from the host processor. Peripherals with bidirectional links also generate and send ECC. Peripherals in unidirectional DSI systems, although they cannot report errors to the host, still take advantage of ECC for correcting single-bit errors in the packet header.

Checksum generation for long packet payloads

The long packet (LP) is protected by checksum over the payload only. The checksum can only indicate the presence of one or more errors in the payload. Unlike ECC, the checksum does not enable error correction. For this reason, checksum calculation is not useful for unidirectional DSI implementations, since the peripheral has no means of reporting errors to the host processor. The details of the algorithms for checksum and ECC can be found in the DSI specification and are not discussed here.

Thus we see that DSI is a protocol that caters to the high-speed need for the display subsytem, at the same time allowing various types of low-speed communication. All care has been taken in the standard to ensure fault tolerance and data integrity. Moreover, a well-defined set of commands and protocols allows various manufacturers of displays and of application processors to arrive at a quick interoperability.

3D DISPLAYS

In today's world 3D movies have become a commonplace experience in theaters. It has also become the norm to recreate the theater experience at home. Any new idea or feature that enhances the experience of watching movies in theaters finally trickles down and finds its way to our living rooms in the form of a new home theater system. Home theater systems have been here for quite a while, and they already provide rich features like Dolby sound, high-definition video, and so on. Adding the 3D capability would take the movie-watching experience at home to the next level.

Stereoscopy overview

Before getting into the details of 3D displays, it is important to understand how the 3D effect is created. Any image comprises a real-life scene or an imaginary scene captured and represented in a 2D way, and this scene is displayed on a 2D display screen. While viewing such an image there is no perception of depth. The process of creating a 3D effect starts with the image creation process. In order to create the perception of depth the same scene is captured as two separate 2D images. Both the images are captured the same way at the same instant of time. The only difference is that the second image is captured with a slight horizontal displacement or offset, which is approximately equal to the offset between the two eyes of humans. These two images can be denoted as the left image and right image. Now by making the left eye of the viewer view only the left image and the right eye of the viewer view the right image, the viewer can perceive depth in the image. This process of creating the perception of depth in the brain of the viewer using two separate 2D images is called *stereoscopy*.

3D display fundamentals

The 3D displays are supposed to create the 3D effect using the concept of stereoscopy. The basic principle of stereoscopy involves presenting a separate image for each eye. However, using a single display screen, both the eyes of the viewer can view both the images. How do we prevent this from happening? This is where the special glasses are used while viewing 3D movies. The glasses use different technologies. Some of the common ones are the polarizer filter based and active shutter based. In the polarized filter-based technology, the left and right images are polarized oppositely, superimposed, and displayed on the screen. The glasses are fitted with appropriate polarizing filters for each eye; these filters allow only those images to pass through that are polarized the same way. In the active shutter system the glasses are fitted with shutters for each eye that can be closed and opened independently: when the left image is displayed on the screen, the shutters for right will be closed and vice versa. This involves synchronization between the glasses and the display. This is the reason why one needs to use special glasses for watching 3D movies.

The role of display interface in 3D displays

The 3D display receives the left and right images through the display interface from a source device and makes the necessary changes (either polarizing them or closing the shutter) before displaying the image. The main role of the display interface is to indicate to the display device whether the image being sent is for left or right eye. This information is very critical for the display device without which the 3D effect cannot be created. All display interfaces can display a certain number of frames or images every second, and this number is equal to the refresh rate of the video mode being used. For a refresh rate of 60 Hz, 60 frames or images are displayed on the screen every second. The same video mode on a 3D display requires 60 images to be sent for the left eye and another 60 images for the right eye every second. This doubles the bandwidth requirement for the display interface. There are means to circumvent this as well. Special timing modes are defined for 3D video content transmission over a display interface. There are different ways in which the left and right image can be transmitted from the source to sink. One way is to double the active region of the frame and send the left image first and then the second image, this is referred to as frame packing structure as per the HDMI specification. Another way is to scale down the image by two on either horizontal or vertical direction and pack both the left and right images within the same active region and this is referred as side-by-side packing or top-and-bottom packing respectively as per the HDMI specification. Usually a Data Island Packet in the case of HDMI is sent from the source to the sink to indicate which type of 3D video timing and packing format is being used. With this information the sink will extract the left and right images accordingly from the pixel stream and display it accordingly. Similar mechanisms exist for other display interfaces like DP etc. as well.

CONCLUSION

In this chapter we quickly glanced over the display interfaces for both external and internal display devices. On any device like smartphone or ultrabook etc. display devices play a major role in determining the user experience and the battery life of the device. This directly translates as innovations in the display interface specifications to deliver more pixels at faster rates with less power.

Multimedia Interfaces

INTRODUCTION

Following up on the general discussion of various interfaces in Chapter 3, this chapter discusses, in detail, the various multimedia interfaces, their applicability to various scenarios, and their capabilities. After this chapter, the reader should understand when to use a particular interface and what advantages they have over other interfaces for a particular design.

The multimedia subsystem primarily covers two things: audio/video playback and audio/video capture. Therefore the multimedia subsystem covers audio, graphics, and imaging devices of a system. As you might guess, a multimedia subsystem generates and consumes a lot of data; therefore, most of the time the audio, graphics, and imaging devices have separate data and control channels. At times, however, the control and data signals are both sent on the same interface on separate logical channels. The data channels are high-speed interfaces while the control channels are low speed. It must be noted that the low-speed interfaces used as control channels for multimedia may actually act as data channels for other subsystems, provided the data transfer rate supported by the interface is good enough. In modern SoCs, typically the interfaces I^2C and SPI are used as the control channel, and various other high-speed interfaces are used for data.

I^2C BUS

The I^2C bus or IIC bus, which is an acronym of inter-IC bus, is one of the most popular interfaces that one can see in any printed circuit board used in multimedia products. I^2C was initially developed by Philips Semiconductor as a simple bidirectional two-wire bus for efficient inter-IC control, and due to its popularity it has become a de facto world standard that is implemented over a wide range of integrated circuits. All I^2C bus-compatible devices incorporate an on-chip interface, which allows the master to communicate to any slave device residing on the bus. Various topological setups are possible, including peer-to-peer and broadcast-type network. But most often a central controller tends to establish a peer-to-peer connection to a device in a network connected in a linear bus fashion. Since a slave and master relationship exists in the I^2C protocol, multimaster topology is allowed if there is a shared resource on the bus.

Some of the features of the I^2C bus include:

- Only two bus lines, a serial data (SDA), and a serial clock (SCL).
- Each device connected to the bus is addressable by a unique address, which can be 7 or 10 bits wide. Thus 1024 unique devices can be connected in an I^2C system.
- Each device can be either master or slave, and relationships exist at all times; masters can operate as master-transmitters or as master-receivers.
- It is a true multimaster bus including collision detection and arbitration to prevent data corruption if two or more masters simultaneously initiate data transfer.
- It is a serial, 8-bit-oriented, bidirectional data transfer protocol. This means transactions can occur in multiples of 8 bits of data.
- The I^2C protocol supports multiple modes of operation, and it can have speeds up to 100 kb/s in *standard-mode*, up to 400 kb/s in *fast-mode*, up to 1 Mb/s in *fast-mode plus*, up to 3.4 Mb/s in *high-speed mode,* or up to 5 Mb/s in *ultra-fast-mode*.
- The number of ICs that can be connected to the same bus is limited only by a maximum bus capacitance.

The I^2C bus protocol

The SDA and SCL lines are used to transfer data between the devices connected onto the bus. Each device has a unique address (which can be 7-bit or 10-bit) and can operate as either a transmitter or receiver, depending on the requirement of the device. A master-slave relationship exists between the devices during the transfer phase. The master always initiates a data transfer on the bus and generates the clock signals to permit that transfer. The addressed device is considered as a slave. Fundamentally, all transactions are initiated by the master. To begin a transaction, the master generates a START condition on the bus and then transmits the address of the slave that it wishes to communicate with, along with an indication of the type of transaction (read/write). The slave, if it is present, sends an acknowledgment to the master. This begins the transfer. After receiving each byte, the receiver acknowledges it. During a remote read operation the sender (master here) waits for this acknowledgment before transmitting the next byte.

The receiver (slave) does not acknowledge the last byte in a transaction—
this indicates to the sender that no more data is to be sent.

Wired and logic

The heart of the I²C operation is the wired and logic and the I/O structure to
monitor the line voltage level as shown in Figure 5.1.

Thus the transmitter and receiver, that is, both the master and slave, continuously monitor the lines to understand the status of transaction going on over
the bus. Note that any device can pull the line low through the MOS switch
as shown in Figure 5.1. Thus the transmitter can verify if the high bit (logic 1)
transmitted by it is available over the bus or not. This mechanism is used for
contention detection (in multimaster), acknowledgment monitoring, and
clock stretching, which are discussed in detail in the following paragraphs.

One clock pulse is required for each data bit transferred. The data on the
SDA line must be stable during the HIGH period of the clock. The HIGH
or LOW state of the data line can only change when the clock signal on

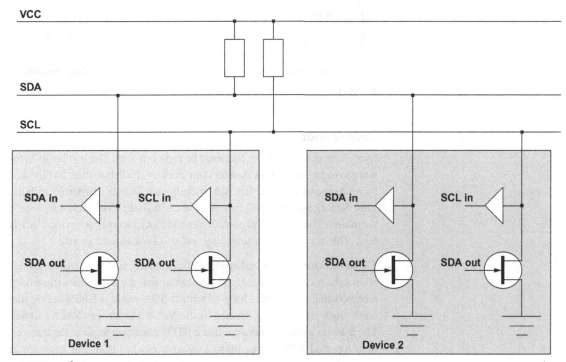

■ **FIGURE 5.1** I²C bus and interface hardware.

130   **CHAPTER 5** Multimedia Interfaces

the SCL line is LOW. Thus, in most of the system the data changes after the falling edge of SCL clock.

START and STOP condition

As shown in Figure 5.2, all transactions must begin with a START (S) and terminate with a STOP (P) condition as defined below:

- A START condition is defined by a HIGH to LOW transition on the SDA line while SCL is HIGH.
- A STOP condition is defined by a LOW to HIGH transition on the SDA line while SCL is HIGH.
- START and STOP conditions are always generated by the master. The bus is considered to be busy after the START condition.
- The bus is considered to be free again a certain time after the STOP condition is available on the line.
- The bus stays busy if a repeated START (Sr) is generated instead of a STOP condition.

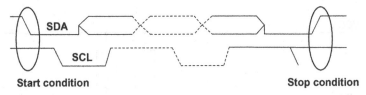

Start condition   **Stop condition**

■ **FIGURE 5.2** START and STOP condition on the bus.

Data format

Every byte put on the SDA line must be eight bits long. The number of bytes that can be transmitted per transfer is unrestricted. Each byte must be followed by an Acknowledge or a Not Acknowledge bit. Data is transferred with the Most Significant Bit (MSB) first. A slave can hold the clock bit low to throttle the master. The master will transfer bits and clock bits only when the clock is high. This is called clock stretching and will be discussed shortly.

The *acknowledge (ACK)* takes place after every byte. The acknowledge bit allows the receiver to signal the transmitter that the byte was successfully received and another byte may be sent. If SDA remains HIGH during this ninth clock pulse, this is defined as the *Not Acknowledge (NACK)* signal. The master can then generate either a STOP condition to abort the transfer, or a repeated START condition to start a new transfer. An example in Figure 5.3 shows a complete transaction.

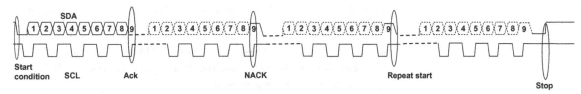

■ **FIGURE 5.3** An example of a complete bus transaction.

Clock stretching

A slow slave device can do a clock stretching to pause a transaction by hold-ing the SCL line LOW. The transaction cannot continue until the line is released HIGH again. On the byte level, a device may be able to receive bytes of data at a fast rate but needs more time to store a received byte or prepare another byte to be transmitted. Slaves can then hold the SCL line LOW after reception and acknowledgment of a byte to force the master into a wait state until the slave is ready for the next byte transfer in a type of handshake procedure. On the bit level, a device (either master or slave) can slow down the bus clock by extending each clock LOW period to match its bit-processing capabilities. This essentially means that even a slow device can be attached to a fast device, but the dynamics of the wired and logic will allow the transaction to occur at that lowest speed among the device connected.

Slave address and R/W bit

This address is 7 (or 10) bits long followed by an eighth bit, which is a data direction bit (R/W):

- A zero indicates a transmission (WRITE).
- A one indicates a request for data (READ).

A data transfer is always terminated by a STOP condition (P) generated by the master. However, if a master still wishes to communicate on the bus, it can generate a repeated START condition (Sr) and address another slave without first generating a STOP condition. After a read request it is the slave's duty to transfer the data byte.

10-bit addressing

To increase the number of devices that can be attached to a bus, a 10-bit addressing is used to expand the number of possible addresses. Devices with 7-bit and 10-bit addresses can be connected to the same I²C bus, and both 7-bit and 10-bit addressing can be used in all bus speed modes. Currently, 10-bit addressing is not being widely used. The 10-bit slave address is formed from the first 2 bytes following a START condition (S) or a repeated START condition (Sr).

■ The first 7 bits of the first byte looks like "1111 0XX" of which the last 2 bits (XX) are the two Most Significant Bits (MSBs) of the 10-bit address that is being generated.
■ The next 8 bits of the second byte define the remaining eight Least Significant Bits (LSBs) of the address.

The format of a 10-bit addressing system is shown in Figure 5.4.

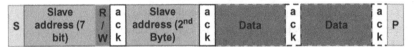

■ **FIGURE 5.4** A master-transmitter addresses a slave-receiver with a 10-bit address.

Thus we see that a specific address value in the 7-bit addressing mode indicates a 10-bit addressing type. Consequently this combination of address cannot be used in any addressing. They are therefore called the reserved address. There are a few more specific addresses that are reserved and are listed in Table 5.1.

Table 5.1 Reserved Addresses

Slave Address	R/W Bit	Description
0000 000	0	General call address
0000 000	1	START byte
0000 001	X	CBUS address
0000 010	X	Reserved for different bus format
0000 011	X	Reserved for future purposes
0000 1XX	X	HS-mode master code
1111 1XX	1	Device ID
1111 0XX	X	10-bit slave addressing

Notes: *Note that if it is known that the reserved address is never going to be used for its intended purpose, a reserved address can be used for a slave address.*

General call address

The general call address (all 8 Address bit is Zero) is for addressing every device connected to the I^2C bus at the same time. However, if a device does not need any of the data supplied within the general call structure, it can ignore this by NACK. If a device does require data from a general call address, it acknowledges this address and behaves as a slave-receiver.

There are two types of general call:

- When the least significant bit B is a zero.
    - The second byte has 0000 0110 (06h): Reset and write the programmable part of the slave address. On receiving this 2-byte sequence, all devices designed to respond to the general call address reset and take in the programmable part of their address.
    - The second byte has 0000 0100 (04h): Write the programmable part of the slave address by hardware.
    - Behaves as above, but the device does not reset.
- When the least significant bit B is a one.
    - When bit B is a one, the 2-byte sequence is ignored.

Multimaster mode

Multimaster bus configuration is also possible. This means more than one master could try to initiate a data transfer at the same time. An arbitration procedure has been developed to avoid any data loss. This procedure relies on the wired-AND connection of all I²C interfaces to the I²C bus. If two or more masters try to put information onto the bus, the first to output a one when the other produces a zero loses the arbitration. Generation of clock signals on the I²C bus is always the responsibility of master devices. During the arbitration phase, only the data line is monitored to define the arbitration loss. The device losing the arbitration will only monitor the I²C signals until a STOP is detected, after which it may try again for data transfer. The process of arbitration with wired and logic defines the type of I/O the I²C device must have. The whole process is shown in Figure 5.5.

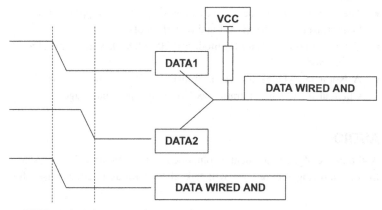

■ **FIGURE 5.5** Data arbitration in multimaster configuration.

Bus speeds

The I²C specification provides five operating speed categories:

- Standard-mode (Sm), bit rate up to 100 kb/s
- Fast-mode (Fm), bit rate up to 400 kb/s
- Fast-mode plus (Fm+), bit rate up to 1 Mb/s
- High-speed mode (Hs-mode), bit rate up to 3.4 Mb/s
- Ultra-fast-mode, bit rate up to 5 Mb/s

Out of these the ultra-fast mode is only unidirectional and is not backward compatible.

Fast-mode and fast-mode plus

Except for the speed of 400 kb/s, fast-mode devices do not have any specific protocol requirement. The requirement comes mostly from the physical layer I/O design. The bus capacitance of less than 200 pF is acceptable. Moreover, instead of a pull-up resistance, a 3-mA current source is recommended.

High-speed mode (Hs-mode)

Hs-mode devices have speed up to a bit rate of up to 3.4 Mb/s, and they are fully downward compatible with fast-mode+, fast-, or standard-mode (F/S) devices for bidirectional communication in a mixed-speed bus system.

There are a few changes as listed below:

- No arbitration or clock synchronization is performed during Hs-mode transfer in multimaster systems, which speeds up bit-handling capabilities.
- Hs-mode master devices generate a SCL signal with a HIGH to LOW ratio of 1 to 2. This relieves the timing requirements for setup and hold times.
- The inputs of Hs-mode devices incorporate spike suppression and a Schmitt trigger at the SDAH and SCLH inputs.
- The 8-bit master code for Hs-mode is (0000 1XXX), which is transferred in F/S mode.
- A Not-acknowledge bit (A) is transferred during the address phase, which also helps in arbitration as NACK is a high bus stage.

AUDIO

Audio is one of the fundamental components of the multimedia subsystem. In the following sections we will talk about various interfaces used for hosting the audio device. The top five interfaces used for audio devices are I2S, Micro-wire, SPDIF, SLIMbus, and USB. The USB interface is generic and used across multiple subsystems.

I2S (inter IC sound) bus

Many digital audio systems are being introduced into the consumer audio market, including compact disc, digital audio tape, digital sound processors, and digital TV sound. The digital audio signals in these systems are being processed by a number of very-large-scale integration (VLSI) application-specific integrated circuits (ASICs) for

- A/D and D/A converters
- digital signal processors
- error correction for compact disc and digital recording
- digital filters
- digital input/output interfaces

Standardized communication structures are vital for both the equipment and the IC manufacturer because they increase system flexibility. To this end, the inter-IC sound (I2S) bus, a serial link especially for digital audio, was developed.

Basic serial bus requirements

The I2S bus handles only the audio data, while the other signals, such as subcoding and control, are transferred separately. To minimize the number of pins required and to keep wiring simple, a three-wire serial bus is used consisting of

- A data line for two time-multiplexed data channels
- A word select line
- A clock line

Since the transmitter and receiver have the same clock signal for data transmission, the transmitter, as the master, has to generate the bit clock, word-select signal, and data. In complex systems, however, there may be several transmitters and receivers, which makes it difficult to define the master. In such systems, there is usually a system master controlling digital audio data-flow between the various ICs. Transmitters then have to generate data under the control of an external clock and so act as a slave (Figure 5.6).

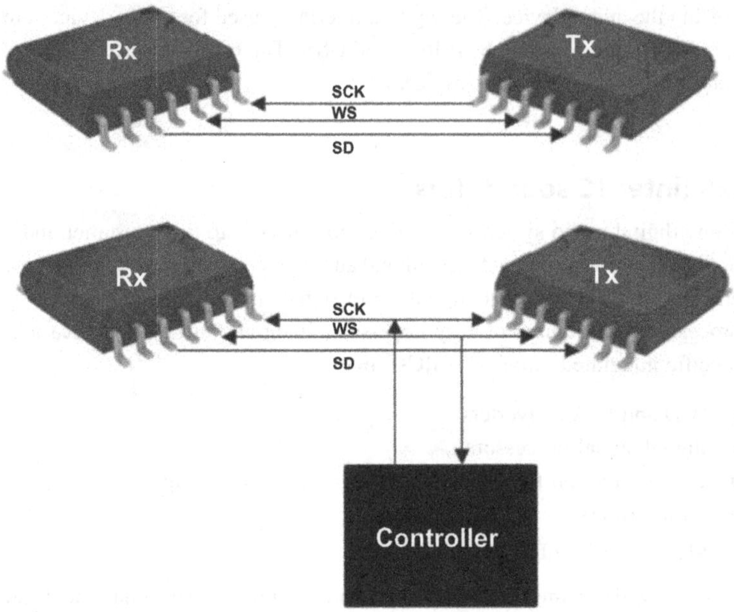

■ **FIGURE 5.6** Clock directions for various configurations of master.

As shown in Figure 5.6, the bus has three lines:

■ continuous SCK
■ word select (WS)
■ SD

SD. SD is transmitted in a complement of two's with the MSB first. The MSB is transmitted first because the transmitter and receiver may have different word lengths. It is not necessary for the transmitter to know how many bits the receiver can handle, nor does the receiver need to know how many bits are being transmitted. When the system word length is greater than the transmitter word length, the word is truncated (least significant data bits are set to zero) for data transmission. When the receiver receives more bits than its word length, the bits after the LSB are ignored. On the other hand, if the receiver is sent fewer bits than its word length, the missing bits are set to zero internally. And so the MSB has a fixed position, whereas the position of the LSB depends on the word length. SDA sent by the transmitter may be synchronized with either the trailing (HIGH-to-LOW) or the leading (LOW-to-HIGH) edge of the clock signal. However, the SDA must be latched into the receiver on the leading edge of the SCL signal, and so there are some restrictions when transmitting data that is synchronized with the leading edge.

Word select. The word select line indicates the channel being transmitted:

- WS = 0; channel 1 (left)
- WS = 1; channel 2 (right)

WS may change either on a trailing or leading edge of the SCL, but it does not need to be symmetrical. In the slave, this signal is latched on the leading edge of the clock signal. The WS line changes one clock period before the MSB is transmitted. This allows the slave transmitter to derive synchronous timing of the SDA that will be set up for transmission. Furthermore, it enables the receiver to store the previous word and clear the input for the next word.

Figure 5.7 shows the relationship between the clock, word select, and the SDA.

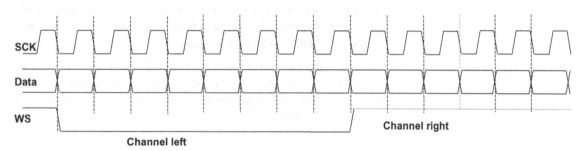

■ **FIGURE 5.7** The relationship between the clock, word select, and the serial data.

Timing

In the I2S format, any device can act as the system master by providing the necessary clock signals. A slave will usually derive its internal clock signal taking into account the propagation delays between master clocks and the data and/or word-select signals. These delays are as a result of

- The delay between the external (master) clock and the slave's internal clock.
- The delay between the internal clock and the data and/or word-select signals.

For data and word-select inputs, the external to internal clock delay is of no consequence because it only lengthens the effective setup time.

The major part of the time margin is to accommodate the difference between the propagation delay of the transmitter, and the time required to set up the receiver. All timing requirements are specified relative to the clock period or to the minimum allowed clock period of a device. This means that higher data rates can be used. Many ASIC uses speed up to 24 MHz of clock signal.

Micro-wire serial interface

Micro-wire, often spelled μ-Wire, is a simple three-wire serial communication interface. In essence it is a subset of SPI: It is full duplex and uses SPI mode 0 as defined in the TI (Texas Instrument) specification of SPI. Micro-wire chips tend to need a slower clock in comparison to newer SPI versions; for example, 2 MHz in comparison to 20 MHz in the case of SPI. Some Micro-wire chips also support a four-wire mode.

M-Wire+ (Micro-wire/Plus) is an enhancement of Micro-wire and features full-duplex communication and support for SPI modes 0 and 1. There was no specified improvement in SCL speed.

Here, the microcontroller in the system is usually called the master and the controlled device is called the slave. Micro-wire can connect one master to many slaves. However, its performance parameters make it better suited to one slave per master.

Protocol description and features

Micro-wire is a simple three-wire serial communications protocol that handles serial communications between a controller and peripheral devices. The signal SI (SERIAL IN) carries input to the microcontroller, SO (SERIAL OUT) carries output from the microcontroller, and SK (SERIAL CLOCK) carries a clock signal sourced from the master. Moving data onto or off of a wire is called "shifting in and shifting out." The clock signal is like a square wave due to its low speed of operation, and the distortion is minimal. Data is shifted out on SK's falling edge and shifted in on SK's rising edge. This is shown in Figure 5.8.

Although Micro-wire is a type of SPI, it actually pre-dates the definition of SPI standards. SPI has a wider number of variations. For example, some SPIs shift data out on SK's rising edge and shift it in on the falling edge.

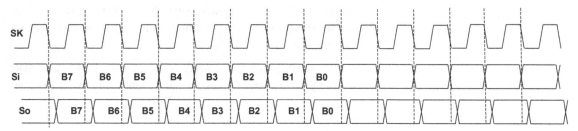

■ **FIGURE 5.8** Micro-wire bus protocol.

System interface

The system usage is straightforward. When two or more peripherals are connected to a Chip Select signal, it is essential to choose the desired peripheral.

There can be two slaves connected to the master. SI, SO, and SK signals from the master are connected to both the slaves. The chip selection of each slave is uniquely driven by the master.

A typical timing diagram shown in Figure 5.9 comprises a CS signal going high or low (depending on implementation) while the clock signal is still

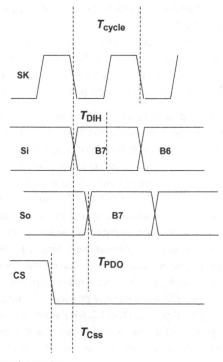

■ **FIGURE 5.9** A typical timing diagram.

low. The clock edges have sufficient margin for sampling the valid data in. The timing parameter are May vary from chip to chip but most popular around clock speed of 1-10 MHz.

Micro-wire interfaces are used mostly with serial E2PROMs and ADC (analog-to-digital converter) chips on microphones, I/O expanders, and so on.

SLIMbus

The *Serial Low-power Inter-chip Media Bus (SLIMbus)* is an MIPI Alliance standard interface between application processors and peripheral components in mobile devices. SLIMbus is one of the newly defined MIPI protocols mostly adopted for interfacing audio devices with a system, but this protocol is well suited for varied type of data transfer required in any multiprocessor system. *SLIMbus* tries to simplify the various legacy low-speed interfaces that present a serious challenge to SoC manufacturers in terms of their interoperability and other system engineering requirements. SLIMbus attempts to standardize many of the low-speed legacy interfaces. SLIMbus had been successful in mostly audio devices like digital mic and codecs, whereas other applications are also picking up. SLIMBus is a two-pin protocol. The two pins required are the data line (DATA) and the clock line (CLK), which is used to interconnect multiple SLIMbus devices in a multi-drop configuration like I^2C. It also inherently tried to converge various legacy interfaces available to date for low-power SoCs into one universal interface. An attempt has been made here to give a brief overview of the general terminologies used in the SLIMbus standard and a brief introduction to the protocol.

Figure 5.10 shows a typical connectivity diagram for an audio subsystem. Comm Processor and App Processor in the figure refer to the communication processor (such as a 3G modem or Wi-Fi) and application processor, respectively. There is an IPC (interprocess communication) interface required between the two processors. In the example given earlier, control signaling between COMM PROC, APPS PROC, and the audio codec happens via an I^2C interface. For audio data transfer PCM and I2S interfaces are used. There are so many interfaces required for the audio subsystem because none of the existing interfaces can support everything required for the audio subsystem to work. Legacy interfaces could not converge for audio need for many reasons. For example, I^2C is a command-only protocol and does not carry any audio stream. I2S is a data-only protocol with no command support and is also not a multipoint protocol. UART is not suited for audio and is also not multipoint. SPI is not good for audio, not a real multipoint. AC'link is flexible enough in some sense but is not a real multipoint.

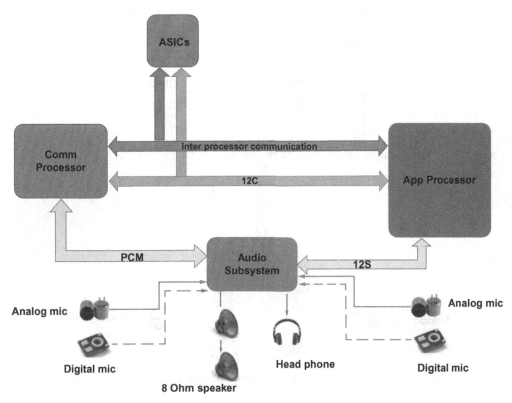

■ **FIGURE 5.10** Audio subsystems with legacy interfaces.

Looking at the complexity of numerous audio and control buses with no convergence in sight, MIPI came up with a standardized and simplified one-for-all interface called SLIMbus. A typical system with a SLIMbus interface for audio systems is shown in Figure 5.11. As can be seen in the figure, the SLIMbus interface combines the IPC, control, and data channels required for the audio subsystem into one.

SLIMbus features

In this section we discuss the fundamentals and features of the SLIMbus interface:

■ Audio, data, and device control all available in a single bus
■ Support for high-quality audio on multiple channels in multi-drop configuration

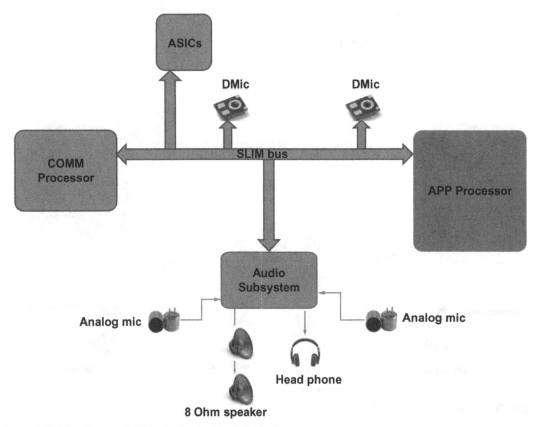

■ **FIGURE 5.11** Audio subsystem with SLIMbus interface showing a unified interface.

- The only two-pin, low-power, multi-drop interface suitable for today's SoCs, conserving real state and power in a PCB (printed circuit board)
- Multiple sample rates on a single bus for various disjoint applications
- Standardized message format for increased interoperability
- On-the-fly clock frequency changes to help in dynamically optimizing bus power consumption
- And many more...

Limitations of existing legacy interfaces for audio, like the limitation of existing legacy buses in terms of the number of channels to be supported and number of devices to be supported, have been resolved to provide greater flexibility and simplicity. The need for separate control interfaces such as I^2C, SPI, Micro-wire, and UART have been eliminated with the use of SLIMbus. SLIMbus also provides greater flexibility in supporting various types of synchronous and asynchronous data at varying rates to be available concurrently within single interfaces. SLIMbus simplifies the system engineering from both the hardware and software perspective.

Figure 5.11 shows how SLIMbus also simplifies the design of PCBs by having a single interface of two wires to host various type of devices that can be placed in a dispersed manner, which otherwise would have required different interfaces to be placed in a constrained manner. It also greatly simplifies the routing of signals on the PCB, allowing greater flexibility in product design. Reduction in pin count for SoCs is another feather in its cap.

On the Physical Layer side, as mentioned earlier, SLIMbus is a two-pin protocol. The two pins required are the data line (DATA) and the clock line (CLK), which are used to interconnect multiple SLIMbus devices. SLIMbus uses a multi-drop bus topology where all bus signals are common to all components on the bus. This reduces bus interconnect wiring in any product using it, while at the same time allowing a multiplicity of devices to be connected to the bus in a multi-drop fashion. A multi-drop configuration requires that only one device transmit data at any given time on the bus to one or more receivers. The SLIMbus transmitter contends for the bus access through the process of arbitration. SLIMbus uses a Time Division Multiplexed (TDM) architecture, which allows multiple receiver and transmitter devices to reside on the bus and allows all devices to intercommunicate within allocated channels and time slots. SLIMbus supports both device-to-device communication and single-device-to-multiple-device communication. SLIMbus is not designed for hot-swap capabilities but may dynamically "drop off" and "reconnect" to the bus while the system is already configured and communication between existing devices is in progress.

SLIMbus devices and device class

In SLIMbus terminology, any device attached to the SLIMbus performing any specified function is called a *component*. A SLIMbus component is made of one or more SLIMbus *devices* and only one SLIMbus *interface device*. Each of the SLIMbus devices in the component implements a specific feature (see Figure 5.12).

A SLIMbus "Device" is a logical implementation of a system function. "Device Class" definitions specify the minimum requirements for Device control information, Device behavior, data transport protocol supported, and data storage necessary to implement a Device of that class. The "Device Class code" identifies the type of Device and the version of the Device

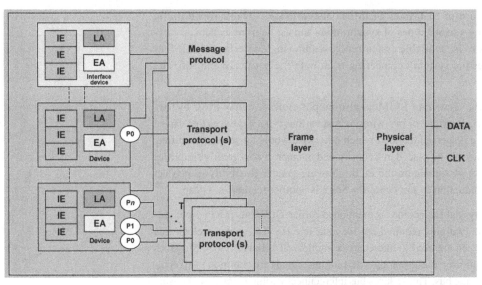

■ **FIGURE 5.12** A typical component model of SLIMbus.

Class. Also specified are the Transport support requirements, such as the number of Ports, its directionality, and transport protocols supported by these Ports. Device Class also provides Message support requirements, identifying which Messages in addition to Core Messages are supported by the Device, the Information support, identifying the Core Information Elements and associated codes that are supported by the Device. The "Operational" requirements identify all other behavior that is important to the operation of a Device that belongs to this Device Class.

There are four SLIMbus Device Classes: "Manager," "Framer," "Interface," and "Generic." These Device Classes permit complete SLIMbus systems to be designed and implemented without any additional Device Classes. Another important definition that is seen in the specifications is "Device Information" and "Value Elements," which apply to all Device Classes.

Information Elements and Value Elements. Information Elements (IEs) and Value Elements (VEs) are data storage elements used to hold status, configuration, or other important information needed by a Device. These IEs and VEs effectively replace registers typically found behind conventional control interfaces such as I^2C or SPI.

An Information Element is specific information that resides in a Device that is available to other Devices via Messages. Various subparts of IEs are: *"Core"* IEs that are the same for all Devices of all Device Classes, *"Device Class-specific"* IEs that are the same for all Devices of a particular Device

Class, but which may be different for all Devices of a different Device Class, and "*User*" IEs that are specific to a particular product or product family. All these subparts have specified memory location as described in the specification.

A *Value Element* provides a standardized method to read and update Device parameters and is typically parameters that are used to configure Device behavior, and region of around 3 kB is allocated for it. The specific message is tightly coupled to the specific VE.

Manager Device. A *Manager Device* in a SLIMbus is the initiator for all administrative transaction, which typically includes enumeration of Components and Devices, bus configuration, channel allocation, and clock configuration. Typically, the Manager would be located in the main application processor. The Manager responsible for administering the bus is known as the active manager. A bus is allowed to have only one active manager at any instance, but it is quite possible to hand over the role of active manager to another manager, if the active manager wants to drop off.

Framer Device. The *Framer* drives a clock signal on the CLK line to all SLIMbus Components and is also responsible for transmitting the Guide and Framing Channel (Framing Information) and communicates with other SLIMbus Devices for synchronization. Framer establishes the Frame Structure on the bus.

Interface Device. The interface Device provides bus management services to the Component in which it resides. It controls the Frame Layer, monitors Message Protocols implemented by the Component, reports information about the status of the Component, and manages the Component reset, and so on. Each Component has one Interface Device.

Generic Device (Function). A Generic Device is generally considered to be the device that provides basic SLIMbus functionality for a Device. For this reason, Generic Devices are also labeled "Function Device." The Generic Device Class is used if no other specific device class for the application functionality exists.

Thus a complete Functional SLIMbus Device (see Figure 5.12) requires the use of a SLIMbus Interface Device, and the associated Enumeration and Logical Addresses (EA and LA), Information and Value Elements (IE and VE), and Ports (P) of each device, which are used to establish bus connections, control and status information flow, and other data flow.

SLIMbus Physical Layer (PHY). As must be clear by now, the DATA and CLK lines are typically attached to two or more SLIMbus Devices contained in one or more SLIMbus Components to form a SLIMbus system. All

SLIMbus Components except one, containing a Framer Device, receive the CLK input, whereas for the Framer, the CLK signal is bidirectional.

For all SLIMbus components, the DATA line is bidirectional and carries data using NRZI encoding. The DATA line is driven on the positive edge and read on the negative edge of CLK. The SLIMbus interface DATA and CLK lines use LVCMOS (Low-Voltage Complementary MOS)-based, single-ended, ground-referenced, voltage mode signaling. In the SLIMbus specification, interface supply voltages of +1.8 and +1.2 V are recommended even though other parts of a SLIMbus Device may use different supply voltages.

Features of LVCMOS (Low-Voltage CMOS) I/Os. Listed here are some of the features of LVCMOS interface derived from the specification, which may typically give an overview of the general directions and requirement of a SLIMbus pins:

- Asymmetrical, unterminated lines separately for data and clock
- SLIMbus I/Os are based on Low-Voltage CMOS technology with slew rate control to limit EMI.
- Maximum bus distance is about 20 cm.
- Total load capacitance limited (15-75 pF).
- Signal slew rate limited for
  - reducing EMI emission and peak current
  - generating clean edges on the CLK line
- "Bus hold function" required for undriven bus.
- No voltage levels are mandated in the specification as different applications will require different logic levels. A parameter set for 1.2 and 1.8 V logic levels is documented. If these levels are used, then the documented parameters are mandatory (normative). Future versions may document other voltage levels.

Basic SLIMbus DATA features

- Data is "half driven" to prevent contentions during driving.
- Data is NRZI coded to allow OR wiring.
- NRZI coding; logical "1" = data line toggles; logical "0" = data line is steady.
- OR wiring allows several devices to talk at the same time without contention and is used for arbitration, acknowledge, and so on.
  The Device writes a logical "1" by toggling the line, while for logical "0" the Device is in Hi-Z state and Data holder logic holds the signal for half the clock cycle. This is shown in Figure 5.13.

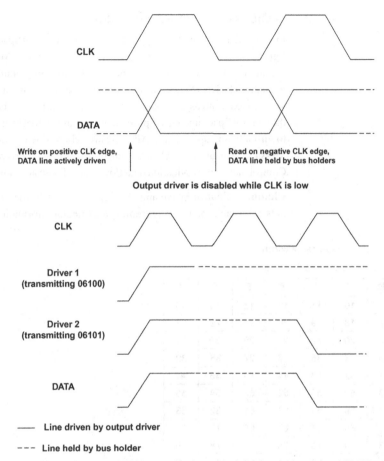

CLK

DATA

Write on positive CLK edge,
DATA line actively driven

Read on negative CLK edge,
DATA line held by bus holders

Output driver is disabled while CLK is low

CLK

Driver 1
(transmitting 06100)

Driver 2
(transmitting 06101)

DATA

—— **Line driven by output driver**

--- **Line held by bus holder**

■ **FIGURE 5.13** Output driver disabled (dashed line) and held by Bus Holder logic while clock is low to achieve a logical OR NRZI signaling.

Basic SLIMbus clock features

■ Clock frequency
  - SLIMbus can accept any kind of clock frequencies from DC to 28 MHz with a bus length of at least 20 cm.
■ Clock Gears
  - For a given reference clock frequency, there are many "Clock Gears" to allow the system designer to match the bus speed and power requirements.
  - Between two consecutive gears, the bus frequency is twice. For example: 12.288-MHz perfectly suits the 48-kHz audio applications; 11.2896-MHz is dedicated to native 44.1-kHz audio applications; 0.768 and 1.536 MHz serve the 8 and 16-kHz voice applications.

SLIMbus operational description

In a SLIMbus system, Data Space channels. The Control Space carries bus configuration, synchronization information, and inter-Device Message communication. The Control Space may be dynamically programmed by the active manager. The Data Space carries application-specific data such as isochronous, synchronous, and asynchronous data streams. The relative bandwidth used by the Control Space and Data Space is configurable so that the bus can be adapted to virtually any application. All these are achieved with Transport Protocols. Messages are used for Control functions. Transport Protocols handle both Control data and Application data flow types. These are shown in Figure 5.14.

Channels. Control Space and Data Space can be further divided into channels, with each channel representing a particular information flow. Channels

Subframe width

0	1	2	3	4	5	6	7
8	9	10	11	12	13	14	15
16	17	18	19	20	21	22	23
24	25	26	27	28	29	30	31
32	33	34	35	36	37	38	39
40	41	42	43	44	45	46	47
48	49	50	51	52	53	54	55
56	57	58	59	60	61	62	63
64	65	66	67	68	69	70	71
72	73	74	75	76	77	78	79
80	81	82	83	84	85	86	87
88	89	90	91	92	93	94	95
96	97	98	99	100	101	102	103
104	105	106	107	108	109	110	111
112	113	114	115	116	117	118	119
120	121	122	123	124	125	126	127
128	129	130	131	132	133	134	135
136	137	138	139	140	141	142	143
144	145	146	147	148	149	150	151
152	153	154	155	156	157	158	159
160	161	162	163	164	165	166	167
168	169	170	171	172	173	174	175
176	177	178	179	180	181	182	183
184	185	186	187	188	189	190	191

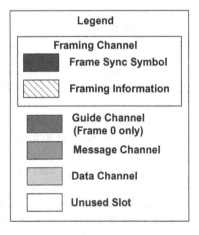

Legend

Framing Channel
Frame Sync Symbol

Framing Information

Guide Channel
(Frame 0 only)

Message Channel

Data Channel

Unused Slot

■ **FIGURE 5.14** A typical channel organization.

are established between a pair of Devices (inter-Device communication), or with many Devices (broadcast communication).

Control Channels. Control Space contains the information needed for a Device to discover the bus state (Framing), to track Messages (Guide), and the Messages themselves (Message). Thus Control Space has three different types of channels: "Framing," "Guide," and "Message." The Framing Channel carries Frame Sync symbol and Framing Information in two Slots of each Frame and carries the bus configuration parameters to properly synchronize. Flow control is not available.

The Guide Channel occupies two Slots (one Slot pair) per Superframe. The Guide Channel is carried in one Slot for the first and second Frame of a Superframe and carries all the necessary information for a Component to acquire and verify Message Synchronization in the Message Channel.

Devices on the bus use the Message Channel to communicate with each other by means of the Message Protocol. The Message Channel carries various types of information including bus configuration, Device control, and Device status. Flow control is implemented by an acknowledgment symbol. Channel width is programmable. A Typical Channel organization is shown in Figure 5.14.

Data Channels and Segments. The active manager allocates the control and the data channels. The number of Data Channels depends on the size of Dataspace and type of data streams carried by the channels. Data Space can contain up to 256 Data Channels. A Data Channel is a stream of one or more contiguous Data Slots that repeats at a fixed interval. The group of contiguous Slots within a Data Channel is known as a Segment. A Segment Window is defined as the range of Slots that a given Data Channel's Segment could possibly be located in. A Data Channel has equidistant spacing throughout the Frame and thus has fixed intervals relative to Superframe boundaries. Data Channels can be viewed as a virtual bus with their own bandwidth and guaranteed latency. A Segment has three fields called the TAG, AUX, and DATA fields, of which the TAG and AUX fields are optional. These fields are organized within the Segment as shown in Figure 5.15.

The TAG bits are used for flow control for the data channel. The auxiliary (AUX) bits carry side information linked to the content of the DATA field.

Device Addressing. SLIMbus uses a 48-bit Enumeration Address (EA) to uniquely identify Devices during enumeration. After enumeration, the active manager assigns a random 8-bit Logical Address (LA) to the Device. The EA incorporates Manufacturer ID, Product Code, Device Index, and Instance Value for a Device. The Device Index code uniquely identifies multiple Devices within a single Component.

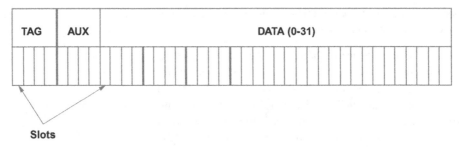

■ **FIGURE 5.15** Segment organization.

Ports. A Port is the part of a Device through which data flows to or from a single Data Channel. A Port provides the connection path to data flow between Devices. A Device may have more than one Port (SLIMbus specification specify the use of maximum of 64 Ports). Port capabilities vary depending on the Device functionality. Typical Port attributes include data directionality, supported Transport Protocols, and the data width.

Figure 5.16 shows a conceptual view of a SLIMbus system with multiple devices. The arrows show the channels established between various devices.

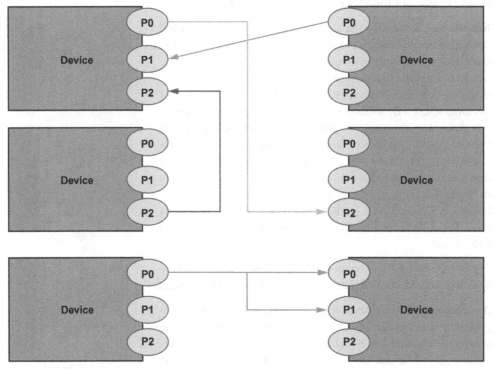

■ **FIGURE 5.16** SLIMbus system showing Data Channel established between multiple devices.

Transport Protocols and Flow Control. A Channel Definition determines the Transport Protocol that will be used in the Data Channel. Many different data types and formats can be simultaneously supported in a SLIMbus system. SLIMbus does not directly support various data formats. Instead, Transport Protocols (including a User-Defined Transport Protocol) are specified, which define data flow type, flow control mechanism, and so on. Data flow between Ports is by one of several Transport Protocols. SLIMbus Device Ports are associated with channels using appropriate channel connection and disconnection Messages. Transport Protocols can be either Uni-cast or Multi-cast types.

The types of Transport Protocols defined in SLIMbus are summarized in Table 5.2.

Table 5.2 SLIMbus-Supported Transport Protocols		
Isochronous (Multi-Cast)	**Pushed (Multi-Cast)**	**Pulled**
Asynchronous—Simplex	Asynchronous—Half duplex	Extended Asynchronous—Simplex
Extended Asynchronous—Half duplex	Locked (Multi-cast)	User Defined 1 & 2

A Data Channel has one data source and can have one or more data sinks depending on the Transport Protocol used. Flow Control used depends on the type of Data involved. TAG bits are used to carry the flow control information. No flow control is needed if the frequency of the CLK line is an integer multiple of the data-flow rate where the Isochronous Transport Protocol can be used. Various types of flow control can be used: single-ended or double-ended flow control. Single-ended data-flow control uses an agreed algorithm (for the Locked Protocol) or uses a "Presence" bit (for the Pulled and Pushed Protocol, respectively).

In a Pushed Protocol, where data rate is equal to or lower than the channel rate, the source Device drives the data flow and the TAG bits indicate the presence of data in the DATA field. In a Pushed Protocol, Data Channel can be connected to multiple sinks (multicast), and the presence TAG bit helps in identifying which slots to sink. Since there is no feedback required from the sinks, it is called Pushed Protocol as it is up to the source when to push the data. When the Pulled Transport Protocol is used, the sink Device requests data from the source Device when needed, and the TAG bits indicate availability of data in the DATA field. With double-ended handshaking, either Device involved in the data transfer can stop or start the transfer using two or more flow control bits in every Data Segment's TAG field.

Frame Structure of SLIMbus. The SLIMbus bit stream uses Synchronous Time Domain Multiplexed structure to define its frame. The Frame Structure has five basic elements: Cells, Slots, Frames, Subframes, and Superframes, shown in Figure 5.17.

- *Cell*: The smallest subdivision of a SLIMbus data flow is the Cell. Each Cell can hold a single bit of information.
- *Slot*: A Slot is defined as four contiguous Cells (4-bits) and is the unit of bandwidth allocation on SLIMbus. Cells within a Slot are labeled C0 through C3 and are transmitted in MSB to LSB order. A variety of Data bit sizes can be formed with these groups of 4-bit Slots.
- *Frame*: A Frame is defined as 192 contiguous Slots. Slots within a Frame are labeled S0 through S191. Slot 0 of each Frame is a Control Space slot that contains the 4-bit Frame Sync symbol. Slot S96 of each Frame is also a Control Space slot that contains 4 bits of Framing Information. A Component uses the Frame Sync data and 32 bits of Framing

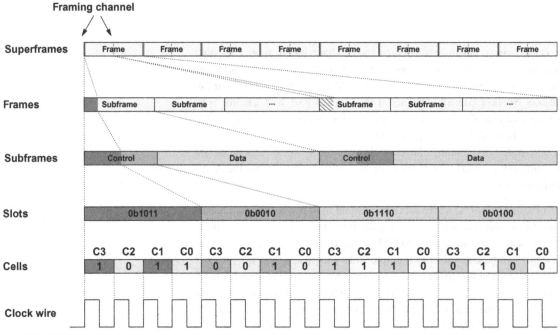

■ **FIGURE 5.17** SLIMbus TDM-based frame structure and its various elements.

Information, from eight successive Frames, called Superframes, to synchronize to the bus. Framing Information also includes Whitening Signal, which is used to reduce the autocorrelation properties of the frame and thereby remove any peaks in the electromagnetic spectrum of the data. Additionally, the Framing Information also contains a 25-Hz timing reference signal, which is the highest common factor of many common audio sample rates. The active Framer writes all Framing Information to the Data line at the appropriate time. Figure 5.17 shows the Frame structure and various elements of the SLIMbus. Figure 5.18 shows how the control and data channels are organized in a frame and how the elements are extracted from the frame.

Subframes: As stated before, SLIMbus Control Space and Data Space information is transported in channels, with each channel representing a particular information flow. The bandwidth used by the Control Space and Data Space is configurable and any configuration is allowed. A Subframe is defined as the division of the Frame Structure at which Control Space and Data Space are interleaved. The first Slot of a Subframe is always allocated to Control Space. Subframes do not have a single, fixed length and is programmable to 6, 8, 24, or 32 contiguous Slots (24, 32, 96, or 128 Cells). Frame Sync and Framing Information only occupy two slots per Frame or the first slot in two different Subframes in a Frame. This means that the first slots in the remaining Subframes of a Frame are used to transmit the other Control Space information, for example, Guide Channel and Message Channel. Any Slots not allocated to Control Space are considered Data Space. One possible example within a single Frame is shown in Figure 5.14.

Superframes: A Superframe is defined as 8 contiguous Frames (1536 Slots). Frames within Superframes are labeled Frame 0 through Frame 7. Each Frame of a Superframe contains the Frame Sync symbol in Slot 0. The first Frame (Frame 0) of a Superframe contains the first 4 bits of a total of 32 bits of Framing Information in Slot 96. Frames 1 through Frame 7 successively also contain 4 bits of Framing Information in Slot 96 with Frame 7 carrying the last 4 bits of Framing Information. The Superframe Sync pattern is transmitted 1 bit at a time in 5 successive Frames. A Component uses a complete set of Framing Information (32 bits over 8 Frames, 4 bits per Frame) and Superframe Sync to achieve Superframe synchronization. The Guide Channel (used for Message Synchronization) consists of 2 slots, one carried in the first Frame of a Superframe and one carried in the second Frame of a Superframe. Figure 5.14 shows a Possible Bus Configuration Arrangement of Control Space and Data Space in a Frame.

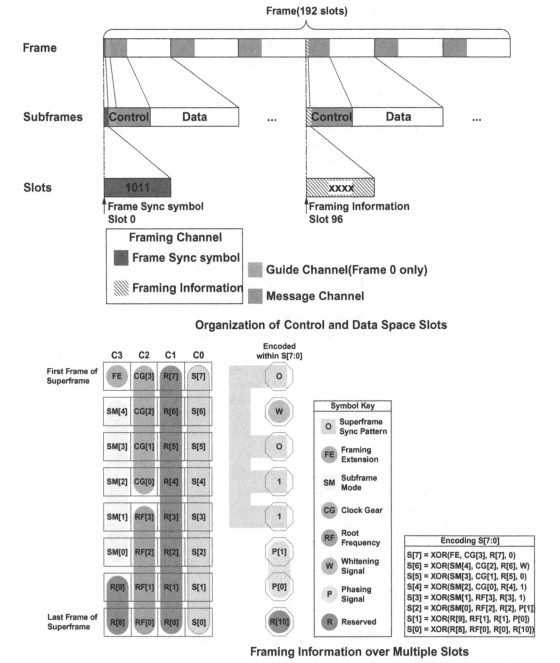

Organization of Control and Data Space Slots

Framing Information over Multiple Slots

■ **FIGURE 5.18** Showing distribution of various elements within a subframe and how framing information is extracted.

SLIMbus Clock Frequencies and Gears. The SLIMbus specification does not specify any absolute SLIMbus CLK frequency. Instead, three frequency definitions are provided: Root, Natural, and Cardinal. Also defined is the Clock Gear concept, which provides 10 Gears for altering the clock frequency by powers of two.

Root Frequency: The Root Frequency is defined as $2^{(10-G)}$ times the frequency of the CLK line, where G is the current Clock Gear. In Gear 10, the CLK frequency is the same as the Root Frequency. The Root Frequency may be a Natural or a Cardinal Frequency. The Root Frequency may be any frequency up to 28 MHz. The Root Frequency may be changed while the bus is active while leaving the Frame Structure unchanged, thus allowing scaling the bus power consumption according to the application.

Natural Frequencies: A Natural Frequency is defined as a CLK frequency that allows a family of isochronous data flows to be supported without flow control, thereby simplifying channel allocation on SLIMbus. For example, Natural Frequencies for 11.025 and 44.1-kHz digital audio sample rate flows include 5.6448, 11.2896, and 22.5792 MHz. Similarly, Natural Frequencies for 8 and 48-kHz digital audio sample rate flows include 6.144, 12.288, and 24.576 MHz. The different sample rate flows can be optimally supported by changing a particular Natural Frequency by 0.5X or 2X using the Clock Gear.

Cardinal Frequencies: In audio applications, one important family of sample rates is comprised entirely of multiples of 4 kHz, that is, 8, 12, 16, 24, 32, 48, 96 kHz, and so on. Another sample rate family is comprised entirely of multiples of 11.025 kHz, that is, 11.025, 22.05, 44.1, 88.2 kHz, and so on. This rate may have non-integer frequency relationships with the natural frequency. In these cases, the CLK line cannot be set to a frequency that is a Natural Frequency for all flows on the bus. The CLK line frequencies 24.576, 12.288, 6.144 MHz, and so forth have special significance because they can carry the 4-kHz family of flows isochronously and the 11.025-kHz family of flows with relatively high efficiency (using push or pull flow control techniques). For this reason, these clock frequencies are referred to as Cardinal Frequencies. There is only one Cardinal Frequency in each SLIMbus Clock Gear.

Clock Gears: Clock Gears provide a range of power-of-two frequency steps for operating SLIMbus. There are 10 Clock Gears, Gear 1 to Gear 10, providing a frequency span of 512. Clock Gears provide a range of power-of-two frequency steps for operating SLIMbus without changing the Root Frequency. Table 5.3 shows the Bus frequency in various Gears.

Table 5.3 Bandwidth Allocation for Various Clock Gears

Gears	Bus Clock (MHz)	Bus BW (Mbps)	Control Space BW (Mbps)		
			1/32	4/32	8/32
10	24.576	24.576	0.768	3.072	6.144
9	12.288	12.288	0.384	1.536	3.072
8	6.144	6.144	0.192	0.768	1.536
7	3.072	3.072	0.096	0.384	0.768
6	1.536	1.536	0.048	0.192	0.384
5	0.768	0.768	0.024	0.096	0.192

Table 5.4 shows Bandwidth required for various audio applications.

Table 5.4 Bandwidth Required for Various Audio Applications

Channel	Application	BW (Mbps)
8 kHz—16 bits	Voice	0.128
16 kHz—16 bits	High-quality voice	0.256
48 kHz—16 bits	Standard audio	0.768
48 kHz—20 bits	Standard audio with S/PDIF aux bits	0.960
48 kHz—24 bits	High-resolution audio	1.152
96 kHz—24 bits	High-resolution audio	2.304
96 kHz—28 bits	High-resolution audio with S/PDIF aux bits	2.688

SLIMbus Messaging. Messaging plays a very important role in SLIMbus, providing an interoperable and standardized mechanism for data and control space configuration. We will be discussing in brief some of these messages. Basic types of Messages are

- Core Messages
    - Device Management Messages
    - Data Channel Management Messages
    - Information Management Messages
    - Reconfiguration Messages
    - Value Management Messages

- Destination-referred Device Class-specific Message
- Destination-referred User Message
- Source-referred Device Class-specific Message
- Source-referred User Message

An example shown in Figure 5.14 shows the position of a typical message channel.

Messages are carried in Message Channels in the Control Space. The Guide Channel (used for Message Synchronization) occupies two Slots per Superframe for all bus configurations. Control Space Slots not used for the Framing Channel or the Guide Channel are available for the Message Channel. The size of the Message Channel varies according to the bus configuration. Messages that are exchanged between Devices consist of four fields: Arbitration, Header, Payload, and Integrity/Response. The Arbitration field varies in size (2 or 7 bytes) depending on the Source Address of the Device requesting arbitration, which could be either an 8-bit Logical Address (LA) or a 48-bit Enumeration Address (EA). The Header field is present in all Messages and varies in size (3, 4, or 9 bytes) as it may use either no address or the Destination Device LA or EA. If all Devices receive a Broadcast Message, then the LA or the EA is not used. The maximum length of the Message Payload field varies between 22 and 28 bytes. The Integrity/Response field is a fixed length of 4 bytes. The Primary and Message Integrity Fields provide full cyclic redundancy check (CRC) coverage of the Message contents. The maximum Message length is 39 bytes.

Message Flow During a Typical Boot. A Component joins the bus by following its appropriate boot process. SLIMbus protocol allows a Component to drop off from the bus while the SLIMbus is active, and rejoin at a later time. The Clock Source Component boots SLIMbus by a specific sequence of operations for starting CLK and for writing the information in the Framing Channel and Guide Channel on the DATA line. During the boot process, the Clock Source Component moves through five different states: Undefined, Checking Data Line, Starting Clock, Starting Data, and Operational. On reaching the Operational state, the Clock Source Component successfully provides all the necessary Framing Information on the DATA line and configures the bus in terms of CLK Frequency and Subframe mode. Lastly, it will also report its presence (REPORT_PRESENT Message) on the Bus via the Message Channel. Any other Device within a Component on the bus that is not a Framer is, by definition, a Clock Receiver Component, which has its own bus boot process. During the boot sequence, a Clock Receiver Component extracts any necessary information about the current state of the bus by monitoring the DATA and CLK lines. During the boot process, Clock Receiver

Components move through Undefined, Reset, SeekingFrameSync, Seeking-SuperframeSync, SeekingMessageSync, and finally, Operational state as shown in Figure 5.19. Once the Operational state is achieved, a REPORT_-PRESENT message is sent and the Clock Receiver Component has all of the Framing Information about the Bus Configuration, has access to the Message Channel, and is now able to perform its function when required.

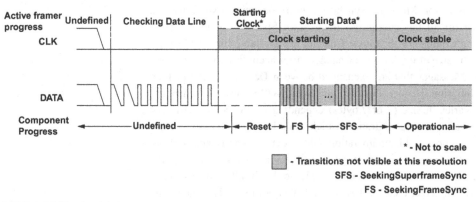

■ **FIGURE 5.19** Typical SLIMbus boot behavior.

Devices become Operational after Bus Boot by sending REPORT_ PRESENT Messages to the active manager, which identifies the Device Class code and version and the Enumeration Address of the reporting Device. The active manager to assign a Logical Address and the Device is considered to be in the Enumerated state. Various state transitions during a typical Boot are shown in Figure 5.20.

Figure 5.21 shows two examples of bus management sequence for RECONFIGURATION by control messages. The first example shows a sequence where the bus management is done for all the devices via broadcast messages. Next figure shows a sequence where the bus management is applied to a particular device on the bus.

Conclusion

SLIMbus is a highly configurable multi-drop bus structure able to support many components simultaneously. In addition, there are powerful messaging constructs to set up and manage data flow between components on the bus. SLIMbus also provides the ability to reconfigure the bus operational characteristics on the fly to adapt to particular system application needs

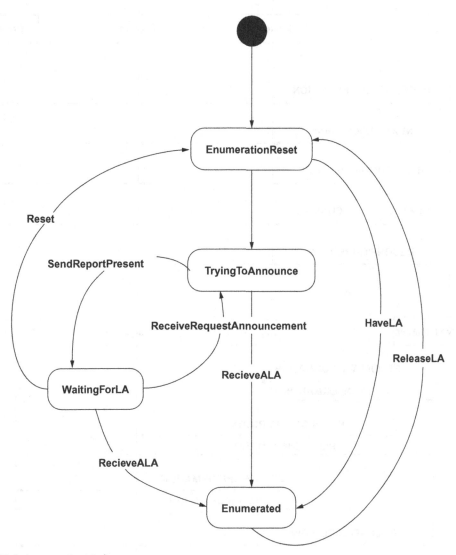

■ **FIGURE 5.20** Device enumeration state diagram.

at runtime. Unlike traditional digital audio bus structures, SLIMbus has the ability to simultaneously and efficiently carry multiple digital audio data streams at widely differing sample rates and bit widths. When using existing digital audio interfaces (PCM, I2S, SSI, AC-97), adding functions and digital audio channels to mobile terminals beyond those for voice communication and simple stereo music applications is very difficult without

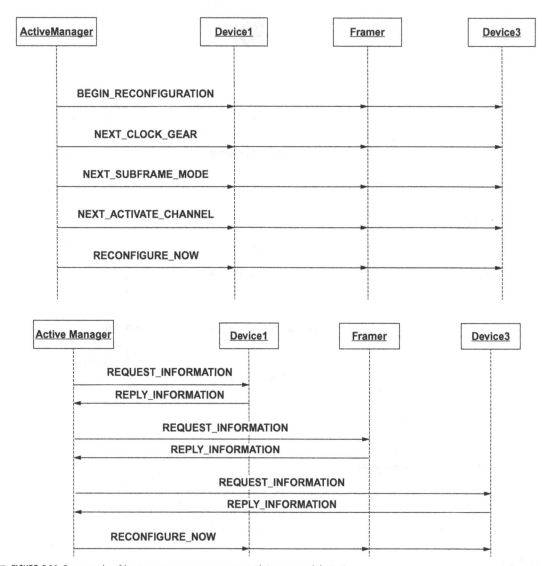

■ **FIGURE 5.21** Two examples of bus management message sequence with its source and destination.

increasing the number of bus structures in the mobile terminal. This is because these legacy interfaces are primarily point-to-point (peer-to-peer) connections with limited channel capacity, and any new device in the system requires its own interface connection. Though legacy interface systems are scalable just by duplicating interface structures, this approach limits design flexibility and is costly in terms of pin count, package size, PCB layout area, and power consumption. SLIMbus provides the mobile terminal industry and other small form factor product makers a standard, robust, scalable, low-power, high-speed, cost-effective, two wire, multi-drop interface that supports a wide range of digital audio and control solutions. As such, it effectively replaces legacy digital audio interfaces such as PCM, I2S, and SSI in such products. SLIMbus can also effectively replace some instances of many digital control buses such as I^2C, SPI, or UART and GPIO, in a mobile terminal or portable product by providing flexible and dynamic assignment of bus bandwidth between digital audio and non-audio control and data functions. SLIMbus is not backward compatible with any current digital audio bus or digital control bus.

Implementing the SLIMbus standard greatly increases the flexibility for designers to realize multiple products within a product line quickly, each with widely varying digital audio and user interface features. This can all be done without the duplication of multiple bus structures within the product. SLIMbus reduces the time-to-market and design cost of mobile terminals and other portable devices by simplifying the interconnection of various functional products from different manufacturers (Table 5.4) with different applications.

As of today, the SLIMbus interface has not achieved popularity or market penetration. The primary reason is that the logic is complex and to implement its design requires a higher number of logic gates.

Universal Serial Bus (USB)

USB, or Universal Serial Bus, which is quite prevalent in present-day PCs and other electronic devices, is the outcome of interface specifications defined by the USB organization (www.usb.org) for high-speed wired communication between electronics systems peripherals and PC/computer or like devices. The USB specification was originally developed in 1995 by many of the industry leading companies.

The initial goal of USB was to define an external PC expansion bus to add peripherals in a manner requiring no or minimal user intervention. The architecture of USB mainly tried to cover the

- controller hardware and software
- robust connectors and cable assemblies
- protocols highlighted as master-slave relationship (called Host-Device in USB terminology)
- expandable to multi-port through the use of hubs

USB offers users a simple universal connectivity and eliminates the mix of different connectors for different devices and other peripherals and removed the connector incompatibility prevalent in the pre-USB era. That means many peripherals can now be connected using a single standardized interface socket. Another main advantage with USB devices is the plug and play. A manual setting is not required to work on the system. It also supports various kinds of data speed, from low-speed mouse to medium-speed compressed audio, and from video to high-speed data transfers from storage devices.

USB introduced "hot plugin," which means that the devices can be plugged in and unplugged without rebooting or turning off the device; that is, when plugged in, everything configures automatically and the user need not set system configurations, such as IRQs, port addresses, etc., and reboot the computer. Once the device is no longer required, the user can simply unplug the cable; the host will detect its absence and automatically unload the driver, making USB a true "plug-and-play" interface. The loading and unloading of the appropriate driver software is done by identification of (PID) Product ID and (VID) Vendor ID. The VID for a manufacturer is supplied by the USB Implementer's forum.

USB is now the most used interface to connect to a bounty of devices loosely termed PC/mobile accessories, which include mouse, hard disk, printer, storage pen, and a lot more. Generally speaking, USB is the most successful interconnect in the history of the electronics industry and personal computing in particular.

The benefits of USB are low cost, expandability, hot-plugging, and outstanding performance, with or without external power, and many more. In short, the user is happy, the industry is happy, and I as an engineer am happy.

Various versions of USB

As USB technology has advanced, new versions of USB have become available from time to time.

USB 1.1. The first USB standard specifies a master/slave interface with a tiered star topology supporting up to 127 devices with up to 6 tiers (each tiered node is called a hub). A PC normally acts as the master (or Host) and all peripherals linked to it act as slaves. The specification tries to minimize the design complexity of the devices by shifting the complexity toward the host. Data transfer rates are defined as Low Speed mode (1.5 Mb/s) and Full Speed mode (12 Mb/s). The maximum length of cable is 5 m. The USB specification allows each device to draw a max of 500 mA of current from the host or hub (limited to 100 mA during startup).

USB 2.0. The 2.0 specification is a superset of 1.1, with the addition of a High Speed (480 Mb/s) data transfer mode. The USB 2.0 protocol also incorporated USB OTG (On-The-Go) where the role of host and the device are dynamically decided based on the requirement. Another subspecification called BC 1.0 (Battery Charging) is incorporated to standardize the battery-charging requirements that evolve with the USB connector's increased use as charging port.

USB 3.0. Released in 2008, USB 3.0 was designed to be backwardly compatible with 2.0 with a socket that fits most combinations of legacy plugs as well as supplying more power (900 mA); it also adds a Super Speed mode of over 4.8 Gb/s data transfer, so it should be able to deliver 600 MB/s after protocol overheads. It is becoming popular for use with external hard disks and other high-speed applications.

USB 3.0 is also a backward-compatible standard with the same plug and play and other capabilities of previous USB technologies. The technology draws from the same architecture of wired USB. In addition, the USB 3.0 specification is optimized for low power and improved protocol efficiency.

It includes a new protocol called PD 1.0 (Power Delivery), which caters to high power requirements and fast charging capability.

USB system overview

The Universal Serial Bus protocol is a host-controlled protocol with only one host per bus and does not support any multimaster bus control. The only exception to this is when a device has been put into "suspend" (a low-power state) by the host, then the device can signal a "remote wakeup." Sometimes, for On-The-Go (OTG)-enabled devices, a Host Negotiation Protocol (HNP) allows two devices to negotiate for the role of host. The USB host is responsible for initiating all transactions and scheduling bandwidths. Data is sent by various transaction methods using a token-based protocol flow controlled by the host (discussed in the subsequent paragraphs of the chapter).

To optimize on the cabling requirement and with envisioning of various accessories usage, USB uses a tiered star topology, similar to that of Ethernet, and thus uses hub as shown in Figure 5.23. Multifunctional devices have integrated hubs, and the tier hierarchy may be embedded in a single physical device. For example, today's monitors or keyboards have integrated hubs that allow USB mouse or external touch screens to be connected. Apart from "daisy chaining" of cables, a tiered star topology has other advantages. The power to each device can be monitored and even switched off without disturbing other connected devices. A hub can selectively connect a low-speed device without bringing down the overall speed of the high-speed device.

Up to 127 devices can be present anywhere in the overall USB tiered bus. This is because the USB is actually an addressable bus system, with a 7-bit address code. To add more devices one simply adds another port/host. A present-day PC has multiple USB ports connected to one or more system host.

A device port that connects to a host is called an upstream port, whereas the port of the hub (which may be referred to as a "root hub" when we talk about a PC hub) that connects to the devices is called the downstream port. An example with a commercial hub is shown in Figure 5.22. USB hubs work transparently as far as the host PC and the device relationship is concerned. A hub must have device functionality for initial communication and configuration to and from the host.

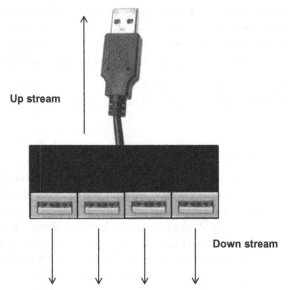

Up stream

Down stream

■ **FIGURE 5.22** Who is UP and who is Down.

The master (host) controls and schedules all communications to the peripherals (devices), which are slaves and respond to commands from the host and may be linked in series through hubs. There is always one root hub, which is always part of the host system. As discussed, a physical device may consist of several logical subdevices that are referred to as *device functions*. A single device may provide several functions built into one physical device. In short, a USB can have two kinds of peripherals: standalone (single function units) or compound devices.

The logical channel connections from the host to the peripheral and vice versa are called *pipes* in USB. A USB device can have 16 upstream pipes and 16 downstream pipes between two nodes. Each logical pipe is unidirectional. Each interface is associated with a single device function and is formed by grouping "endpoints." Talking from network concepts, the hubs are bridges that expand the logical and physical fan-out of the network. A hub has a single upstream connection and can have many downstream connections (Figure 5.23).

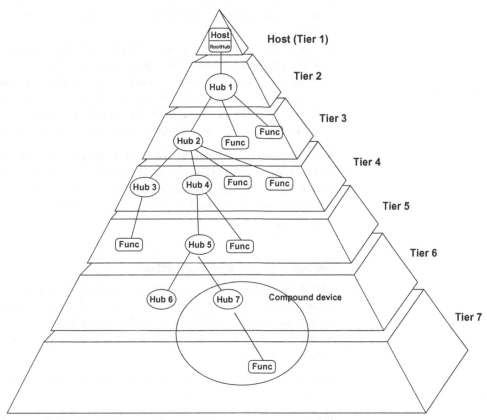

■ **FIGURE 5.23** Illustration of multi-tier hub and functions.

There are two Host Controller Interface Specifications, *UHCI (Universal Host Controller Interface)*, developed by Intel, which has more functionality embedded in software and allows less complex hardware, and the *OHCI (Open Host Controller Interface)*, which has more complex hardware and allows light software. In USB 2.0, a new Host Controller Interface Specification *EHCI (Enhanced Host Controller Interface)* provides one interface standard and thus has only one software implementation.

USB supports four types of data transfers:

- *Control.* Used for sending commands to the device, making inquiries, and configuring the device. This transfer uses the *control pipe*.
- *Interrupt.* Used for sending small burst data that require a guaranteed minimum latency. This transfer uses a *data pipe*.
- *Bulk.* Used for large data transfers that use all available USB bandwidth with no guarantee on transfer speed or latency. This transfer uses a *data pipe*.
- *Isochronous.* Allows data transfer with a reserve and defined amount of bandwidth with guaranteed latency for use in cases in audio or video applications, where line condition may cause loss of data or frames to drop. Because of this guaranteed latency and guaranteed bus bandwidth, it lacks error correction, and there is no halting or retransmission where packets containing errors are resent. This transfer uses a *data pipe*.

As is evident, USB is a serial bus that uses four shielded wires, of which two are power (+5 V and GND), and two are twisted differential signals (D+, D−). It uses a NRZI (Non Return to Zero Invert) encoding scheme for data transmission with a sync field in its data format to synchronize the host and device clocks.

Connectors and electrical details

USB devices have an upstream connection to the host and have a downstream connection to the device. Upstream and downstream connectors are not mechanically interchangeable, thus illegal loopback connections at hubs such as an upstream port connected to a downstream port or vice versa are not possible.

There are commonly two types of connectors, called type A and type B, which are shown in Figure 5.24.

Type A plugs always face upstream and typically find themselves on hosts and hubs (computer main boards and hubs). Type B plugs always face downstream and consequently type B sockets are found on devices.

USB 2.0 errata introduces mini-USB B connectors to fit into miniature electronic devices such as mobile phones. The On-The-Go specification adds peer-to-peer interchangeable functionality to USB and thus has included a specification for mini-A plugs, mini-A receptacles, and mini-AB receptacles.

Both mini-A and mini-B plugs are approximately 3 by 7 mm in size. Micro-USB plugs have a similar width to mini-USB, but approximately half the thickness, enabling their integration into thinner portable devices.

USB 2.0 jack and plug Type A

USB 2.0 jack and plug Type B

USB 3.0 jack and plug Type A

USB 3.0 jack and plug Type B

USB 2.0 jack and plug Type B mini

USB 2.0 jack and plug Type B micro

USB 3.0 jack and plug Type B micro

■ **FIGURE 5.24** Common USB connectors (Type A and Type B).

Electrical

The essence of electrical characteristics with various definitions is listed here:

- USB uses a differential signaling for data transmission, which is encoded using NRZI and is bit-stuffed to ensure adequate transitions in the data stream for reliable clock recovery.

On low- and full-speed devices,

- A differential "1" (K-State) is transmitted D+ over 2.8 V with a 15-K termination to ground and D− under 0.3 V with a 1.5 K pulled up to 3.6 V.
- A differential "0" (J-State) is a D− greater than 2.8 V and a D+ less than 0.3 V with the same pull down/up resistors.

On a receiver,

- A differential "1" as D+ 200 mV greater than D−
- And a differential "0" as D+ 200 mV less than D−.

For a high-speed device,

- J-State is a differential "1" and K-State is differential "0" (in some sense opposite to that of LS signaling).

- The low-speed/full-speed bus has a characteristic impedance of 90 Ω.
- High speed (480 Mb/s) mode uses 18-mA constant current sources for signaling to reduce noise.

USB transceiver (transmitter plus receiver) will have both differential and single-ended outputs.

- *Single-ended zero or SE0* can be used to signify a device reset if held for more than 10 mS. A SE0 is generated by holding both D− and D+ low (<0.3 V), which indicates a reset, disconnect, or End of Packet.
- *Single-ended one (SE1)*: when D+ and D− are both driven high. This condition is avoided in the USB specification, but nevertheless can occur.
- *Idle*: The idle state is indicated differently by the high-speed mode and the full-/low-speed mode. Under the full-/low-Speed mode, the idle state is indicated with a "J" state, whereas under the high-speed mode, a "D+=D−=0" state is used to indicate the idle state. This is equivalent to the "SE0" state under full/low-speed mode.
- *Resume*: Used to wake a device from a suspend state. This is done by issuing a K-State.
- *Start of Packet (SOP)*: Occurs before the start of any low-speed or full-speed packet when the D+ and D− lines transition from an idle state to a K-State.
- *End of Packet (EOP)*: Occurs at the end of any low-speed or full-speed packet. An EOP occurs when an SE0 state occurs for 2 bit times, followed by a J-State for 1 bit time.
- *Reset*: Occurs when an SE0 state lasts for 10 ms. After an SE0 has occurred for at least 2.5 ms, the device may recognize the reset and begin to enter a reset state.

Table 5.5 shows all the states listed.

NRZI encoding. NRZI (Non Return to Zero Inverted) encoding is a method for mapping a binary signal for transmission over some medium. With this encoding scheme,

- A logic 1 is represented by no change in voltage level.
- A logic 0 is represented by a change in voltage level.

Bits Stuffing: The bit stuffing occurs by inserting a logic 0 (a transition), if there are seven consecutive logic 1s. The stuffing is done after the sixth bit. The purpose of the bit stuffing is for synchronization of the USB hardware's phase-locked loop (PLL). If there are too many logical 1s in the data, then there may not be enough transitions in the NRZI-encoded stream to synchronize from. The receiver on the USB hardware automatically detects this

Table 5.5 USB Communication States

Differential '1'	D+ high, D− low
Differential '0'	D− high, D+ low
Single-ended zero (SE0)	D+ and D− low
Single-ended one (SE1)	D+ and D− high (illegal)
Data J state Low-speed Full-speed	 Differential "0" Differential "1"
Data K state Low-speed Full-speed	 Differential "1" Differential "0"
Idle state Low-speed Full-speed	 D− high, D+ low Differential '0' D+ high, D− low Differential '1'
Resume state	• Data K state • It does so by reversing the polarity of the signal on the data lines for at least 20 ms The signal is completed with a low-speed EOP signal Suspended — Resume — D+ and D−: Idle (J) → K state → Idle (J) state > = 20 ms 2 low speed bit times, 1 low speed bit time
Start of packet (SOP)	Data lines switch from idle to K state
End of packet (EOP)	SE0 for 2 bit times followed by J state for 1 bit time
Disconnect	SE0 for ≥ 2 μs
Connect	Idle for 2.5 μs
Reset	SE0 for ≥ 2.5 μs

Suspend device	• Achieved by not sending anything to the device for 3 ms
	  ○ Normally an SOF packet (at full speed)
	  ○ Or a Keep Alive signal (at low speed) is sent by the host every 1 ms and this is what keeps the device awake
	
	Idle state, 2 bits, 1 bit, Idle state
	D+ and D−
	← Keep alive →
	
	• A suspended device must recognize the resume signal, and also the reset signal

extra bit and disregards it. This extra bit stuffing contributes to the extra overhead on the USB.

Figure 5.25 shows an example of NRZI data with bit stuffing. Notice that in the "Data to Send" stream, there are eight 1s. In the encoded data, after the sixth logic 1, logic 0 is inserted. The seventh and eighth logic 1 then follow after this logic 0.

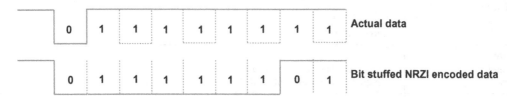

■ **FIGURE 5.25** Signaling vs. data for NRZI encoding without and with bit stuffing.

An example of USB HS communication is shown in Figure 5.26.

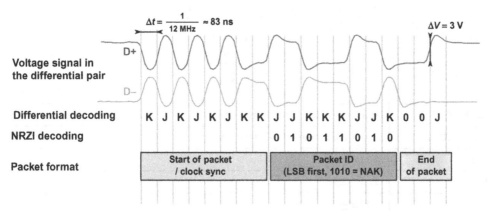

■ **FIGURE 5.26** An example of HS signaling and USB packet.

Speed identification. A 1.5-kΩ pull-up resistor to 3.3 V derived from VBUS connected to D+ D− line helps in speed identification as follows:

- A full-speed device has pull-up connected to D+.
- A low-speed device has pull-up connected to D−.
- A high-speed device will initially present itself as full speed and will do a high-speed chirp during reset and establish a high-speed connection.
- A USB 2.0-compliant downstream-facing device supports all three modes: high speed, full speed, and low speed.

Internal wires. Standardized internal wire colors are used in USB cables, making it easier to identify wires from manufacturer to manufacturer as shown in Figure 5.27. The standard specifies various electrical parameters for the cables.

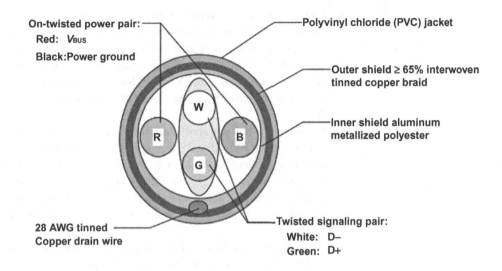

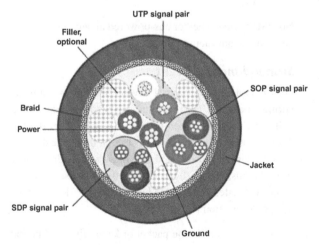

■ **FIGURE 5.27** USB2.0/3.0 internal wire color.

Power (VBUS)

USB is a bus-powered bus; that is, devices obtain their power from the bus. The line supplying the voltage, in this case, is called VBUS.

A USB device specifies its power consumption in the configuration descriptor. A device cannot increase its power consumption to greater than what it specifies during enumeration. There are three classes of USB functions:

- **Low-power bus-powered functions**. Low-power bus-powered functions draw all their power from the VBUS and cannot draw any more than one unit load. The USB specification defines a unit load as 100 mA. Low-power bus-powered functions must also be designed to work down to a VBUS voltage of 4.40 V and up to a maximum voltage of 5.25 V measured at the upstream plug of the device.
- **High-power bus-powered functions**. High-power bus-powered functions draw all their power from the VBUS and cannot draw more than one unit load until they have been configured, after which they can then drain 5 unit loads (500 mA). High-power bus functions must be able to function at a minimum of 4.40 V. When operating at a full unit load, a minimum VBUS of 4.75 V is specified with a maximum of 5.25 V.
- **Self-powered functions**. Self-powered functions may draw up to 1 unit load from the bus and derive the rest of their power from an external source.

No USB device, whether bus-powered or self-powered, can drive the VBUS on its upstream-facing port.

Suspend mode current

For the typical bus-powered device, 500 mA is the maximum load during normal operation. The maximum suspend current is 500 µA. A USB device will enter suspend, when there is no activity on the bus for greater than 3.0 ms. It then has a further 7 ms to shut down the device and draw no more than the suspend current and thus must be only drawing the suspend current from the bus, 10 ms after all bus activity has stopped. In order to maintain connected state, a suspended hub or host, the device must still have its pull-up speed selection resistors enabled.

USB has a start of frame packet or keep-alive sent periodically on the bus. This prevents an idle bus from entering suspend mode in the absence of data.

- A high-speed bus will have micro-frames sent every 125.0 µs ± 62.5 ns.
- A full-speed bus will have a frame sent down each 1.000 ms ± 500 ns.

- A low-speed bus will have a keep-alive that is an EOP (End of Packet) every 1 ms only in the absence of any low-speed data.

Global Suspend is used when the entire USB bus enters suspend mode. However, in *Selective Suspend*, selected devices can be suspended by sending a command to the hub that the device is connected to. The device resumes operation when it receives any non-idle signaling. If a device has remote wakeup enabled, then it may signal to the host to resume from suspend.

Power states. There are various USB states that relate to USB power that a designer needs to know. These states occur during the device power-up and are shown in Figure 5.28.

- *Attached state*: Occurs when a device is attached to a host/hub, but does not give any power to the VBUS line.
- *Default*: Occurs when a device has been reset by the host. At this point, the device does not have a unique device address and has default address 0.
- *Address*: Occurs when a unique address is assigned to it, but it has not yet been configured.
- *Configured*: When functionality of the device has been programmed. At this point, bus-powered devices can draw more than 100 mA.
- *Suspend*: When a device has not seen activity on the bus for 3 ms or more.

USB Endpoints

A device endpoint is a uniquely addressable portion of a USB device that is the source or sink of information between the host and device. All USB devices are required to implement *endpoint 0*, which contains the information about the other endpoints supported by the device. The other endpoints define the various functions of the device. During the *enumeration* process, the endpoint 0 is read by the host. Then the other endpoints can be configured. Control Endpoint or Endpoint 0 is used for communication with the device. Every USB device must support Endpoint 0, which is the default endpoint.

In addition to Endpoint 0, the number of endpoints supported in any particular device is based on its function. Simple functions may need only one IN endpoint. Others may need several data endpoints. The USB specification states a maximum of 16 endpoints for each direction. Data endpoints are bidirectional in general but can be configured to become unidirectional based on device descriptor.

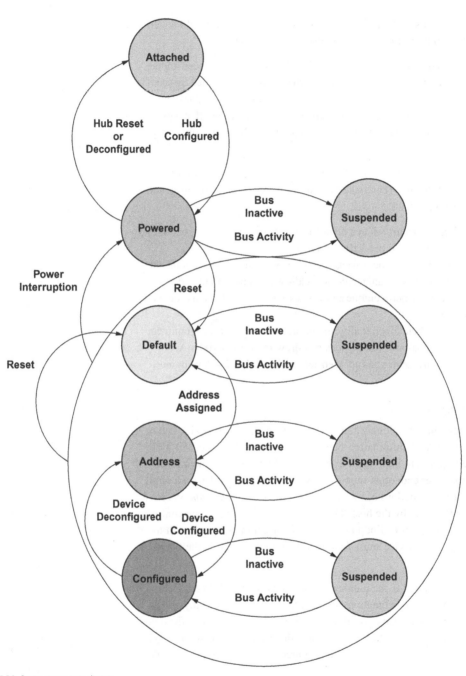

■ **FIGURE 5.28** Device power state diagram.

The USB specification defines four types of endpoints and sets the maximum packet size based on both the type and the supported device speed. The endpoint descriptor is used to identify the type of endpoint and maximum packet size.

The four types of endpoints are

- *Control endpoints*: These endpoint types support control transfers, which all devices must support. Control transfers have a 10% reserved bandwidth on the bus in low- and full-speed devices (20% at high speed).
- *Interrupt endpoints*: These endpoint types support interrupt transfers that use a high-reliability method to communicate a small amount of data. In fact, it is not an interrupt transfer in the true sense but uses a polling that guarantees that the host checks data at a predictable interval. Interrupt transfers have a guaranteed bandwidth of 90% on low- and full-speed devices and 80% on high-speed devices; the bandwidth is shared with isochronous endpoints. Interrupt endpoint maximum packet size is a function of device speed. For HS devices the maximum packet size is 1024 bytes, for Full-speed devices maximum packet size is 64 bytes, and for Low-speed devices maximum packet size is 8 bytes.
- *Bulk endpoints*: Bulk endpoints are commonly used on devices that move relatively large amounts of data using any available bandwidth space. The delivery time varies depending on how much bandwidth on the bus is available and thus is unpredictable. Bulk transactions are resent on errors. The maximum packet size is a function of device speed. High-speed-capable devices support a maximum bulk packet size of 512 bytes. Full-speed-capable devices support a maximum packet size of 64 bytes. Low-speed devices do not support bulk transfer type.
- *Isochronous endpoints*: Isochronous transfers are continuous, real-time transfers that have a pre-negotiated bandwidth and transfer data that have no error recovery mechanism or handshaking. Streaming data (audio or video) uses isochronous endpoints because the occasional missed data is not that important. Isochronous transfers have a guaranteed bandwidth of 90% on low- and full-speed devices (80% on high-speed devices) that is shared with interrupt endpoints. High-speed-capable devices support a maximum packet size of 1024 bytes. Full-speed devices support a maximum packet size of 1023 bytes. Low-speed devices do not support isochronous transfer types.

Pipes. A pipe is a logical connection between the host and endpoints and has a set of parameters associated with them such as bandwidth, transfer type, maximum packet/buffer sizes, and so on. While the device sends and receives data on a pair of endpoints, the client software transfers data through pipes. The default pipe is a bidirectional pipe made up of *endpoint zero IN* and *endpoint zero OUT* with a control transfer type. Figure 5.29 shows this concept of pipes and endpoints.

USB defines two types of pipes:

■ *Stream Pipes* have no defined USB format, and any type of data can be sent down a stream pipe and can retrieve the data out the other end. Data flows sequentially in a predefined direction. Stream pipes support bulk, isochronous, and interrupt transfer types and can be controlled by either the host or device.
■ *Message Pipes* have a defined USB format, and control is initiated by a request sent from the host. Data is then transferred in the desired direction. Message pipes allow data to flow in both directions but support control transfers.

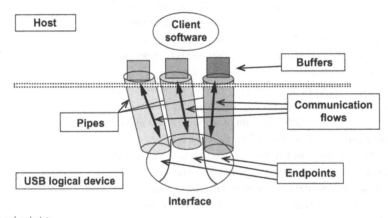

■ **FIGURE 5.29** Pipes and endpoints.

USB Protocol

USB is a host-controlled bus. The host initiates all transactions, which are contained in a series of frames. Each frame consists of a *Start of Frame (SOF)* followed by one or more transactions. Each transaction is made up of a series of packets. A packet is preceded with a *sync pattern* and ends with an *End of Packet (EOP) pattern*. Depending on the transaction, there may be one or more data packets.

USB transaction. Each USB transaction shown in Figure 5.30 consists of packets:

■ *Token packet*: The first packet, also called a token, is generated by the host, which describes what data packet will subsequently follow and whether the data transaction will be a read or write and what the device's address and designated endpoint is, and so on. At a minimum, a transaction has a token packet. Thus the token packet initiates a transaction and identifies the device involved in the transaction. The token packet is always sourced by the host.

- *Data packet*: Contains the main payload. This packet is not always present in a transaction.
- *Status packet*: Used to acknowledge transactions and provides error correction by exchange of handshaking information, reporting if the data or token was received successfully or if the endpoint is stalled or not due to buffer being full. Some transactions may or may not have a handshake packet.
- *Special packets*: There are special packets that facilitate speed differentials. These are sourced by host-to-hub devices.

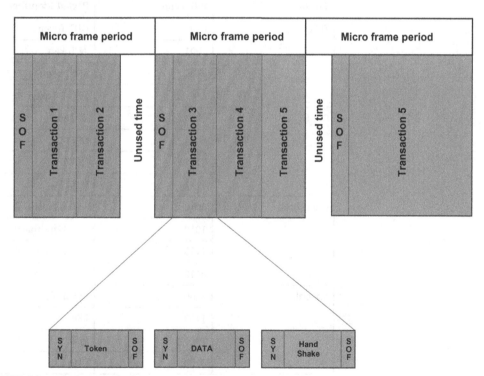

■ **FIGURE 5.30** Token packet in a transaction frame.

Transactions are always placed within frames and are never split across frames.

USB packet fields. USB packets consist of the fields shown in Figure 5.31. Note that the LSB bit is transmitted first.

■ **FIGURE 5.31** Fields in a USB Packet.

- *Sync*. All packets start with a sync field that is 8 bits long for low and full speed or 32 bits long for high speed. It is used for clock synchronization.
- *PID*. The Packet ID field is used to identify the type of packet that is being sent. Table 5.6 shows the possible values.

Table 5.6 PID Values for Different Types of Packet

Group	PID Value	Packet Identifier
Token	0001	OUT Token
	1001	IN Token
	0101	SOF Token
	1101	SETUP Token
Data	0011	DATA0
	1011	DATA1
	0111	DATA2
	1111	MDATA
Handshake	0010	ACK Handshake
	1010	NAK Handshake
	1110	STALL Handshake
	0110	NYET (No Response Yet)
Special	1100	Preamble
	1100	ERR
	1000	Split
	0100	Ping

There are 4 bits to the PID; however, to ensure it is received correctly, the 4 bits are complemented and repeated, making an 8-bit PID in total. Here is the resulting format:

PID_0	PID_1	PID_2	PID_3	$nPID_0$	$nPID_1$	$nPID_2$	$nPID_3$

- *ADDR*. The address field specifies which device the packet is designated for. Being 7 bits in length allows for 127 devices to be supported. Address 0 is not valid, as any device that is not yet assigned an address must respond to packets sent to address zero.
- *Optional Payload Data*. This may be 0-1023 bytes.
- *ENDP*. The endpoint field is made up of 4 bits, allowing 16 possible endpoints. Low-speed devices, however, can only have two additional endpoints on top of the default pipe.
- *CRC* (cyclic redundancy checks). All token packets have a 5-bit CRC while data packets have a 16-bit CRC.
- *EOP* (End of Packet). The EOP is signaled by a single-ended zero (SE0) for approximately 2 bit times followed by a J for 1 bit time.

USB packet description. As described several times earlier, USB has four different packet types. Token packets indicate the type of transaction to follow, data packets contain the payload, handshake packets are used for acknowledging data or reporting errors, and start of frame packets indicate the start of a new frame.

- *Token packets*. There are three types of token packets:
  - In: Indicates to device that host wants to read information.
  - Out: Indicates to device that the host wants to send information.
  - Setup: Used to begin control transfers.

  Token packets have the following format:

Sync	PID	ADDR	ENDP	CRC5	EOP

  Inside an IN, OUT, and SETUP token packet, there is a 7-bit device address, 4-bit endpoint ID, and 5-bit CRC.
- *Start of frame packets*. The start of frame (SOF) packet is sent every 1 ms on a USB full-speed system to help synchronization. On a USB High-Speed system a packet frame is transferred every 125 µs, resulting in 8 packets every 1 ms. These 8 subframes all contain the same frame number.

  The SOF indicates the beginning of a frame and synchronize with the host and is also used to prevent a device from entering suspend mode (if no SOF for 3 ms). SOF packets are only used on full- and high-speed devices and are sent every millisecond as seen in Figure 5.32.

1 Full Speed FS Frame (1ms)							
HS Micro Frame period (125 u sec)	HS Micro Frame period (125 u sec)	HS Micro Frame period (125 u sec)	HS Micro Frame period (125 u sec)	HS Micro Frame period (125 u sec)	HS Micro Frame period (125 u sec)	HS Micro Frame period (125 u sec)	HS Micro Frame period (125 u sec)

■ **FIGURE 5.32** USB SOF in full-speed and high-speed device.

The SOF packet has the structure as shown in Table 5.7. A handshake packet does not occur for an SOF packet.

Table 5.7 The SOF Packet Structure

SYNC	PID	Frame No.	CRC5	EOP
8/32 bit	8 bits	11 bits	5 bits	3 bits

- *Data Packets.* There are two types of data packets: Data0 and Data1. High-speed mode defines another two data PIDs: DATA2 and MDATA. Data packets have the following format:

Sync	PID	Data	CRC16	EOP

  - Maximum data payload size for low-speed devices is 8 bytes.
  - Maximum data payload size for full-speed devices is 1023 bytes.
  - Maximum data payload size for high-speed devices is 1024 bytes.
  The packet ID toggles between DATA0 and DATA1 for each successful data packet transfer. The data toggle acts as additional error detection method. In case of errors the sender updates the data toggle from "1" to "0" but the receiver does not. This sync loss indicates an error. An example of the data toggle in a USB transfer can be seen in Figure 5.33.

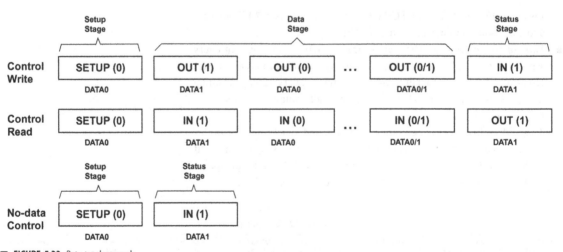

■ **FIGURE 5.33** Data toggle example.

■ *Handshake packets*. There are three types of handshake packets, which contain only PID:

ACK: Acknowledgment

NAK: To indicate that a function is temporarily unable to transmit or receive data. The host never issues a NAK handshake packet to a device.

STALL: A function uses the STALL handshake packet to indicate that it is unable to transmit or receive data. Besides the default control pipe, all of a function's endpoints are in an undefined state after the device issues a STALL handshake packet. The host must never issue a STALL handshake packet. Handshake Packets have the following format:

Sync	PID	EOP

■ *Special packets*. The USB specification defines four special packets.
  - PRE: to indicate that the next packet is low speed.
  - SPLIT: Precedes a token packet to indicate a split transaction. (HS Only)
  - ERR: Returned by a hub to report an error in a split transaction. (HS Only)
  - PING: Checks the status for a Bulk OUT or Control Write after receiving a NYET handshake. (HS Only)

USB transaction types. There are three different transaction types:

■ IN/read/upstream transactions
■ OUT/write/downstream transactions
■ Control transactions

IN/read/upstream transactions. These transactions are initiated by the host by sending an IN token packet. The targeted device responds by sending one or more data packets or a NACK and the host responds with an ACK if transaction is successful.

OUT/write/downstream transactions. The host sends the appropriate token packet (either an OUT or SETUP) and follows with one or more data packets. The receiving device ends the transaction by sending the appropriate handshake packet.

Control transactions. Control transactions identify, configure, and control devices. They enable the host to read information about a device, set the device address, establish configuration, and issue certain commands. A control transfer is always directed to the control endpoint of a device. Control

Setup Token	DATA0 Data (8 bytes)	ACK Handshake	IN | OUT Token	DATA1/0 Data	ACK | NAK | STALL Handshake	OUT | IN Token	DATA1 Data (0 bytes)	ACK | NAK | STALL Handshake

Setup Transaction	Data Transaction	Status Transaction

■ **FIGURE 5.34** Packets during In/Out transaction.

transfers have three stages: the setup stage, the (optional) data stage, and the status stage (see Figure 5.34).

- The setup stage (or setup packet) is only used in a control transaction. The setup stage is always followed by an ACK.
- The data stage is optional in a control transaction. This stage can consist of multiple data transactions and is only required when a data payload exists between the host and device.
- The status stage includes a single IN or OUT transaction that reports on the success or failure of the previous stages. The data packet is always DATA1 (unlike normal IN and OUT transactions that toggle between DATA0 and DATA1) and contains a zero length data packet. The status stage ends with a handshake transaction that is sent by the receiver of the preceding packet.

Control transfer types. There are three types of control transfers: control write, control read, and control no data. Figure 5.35 shows this transfer, and the different packet types used are evident from the figure.

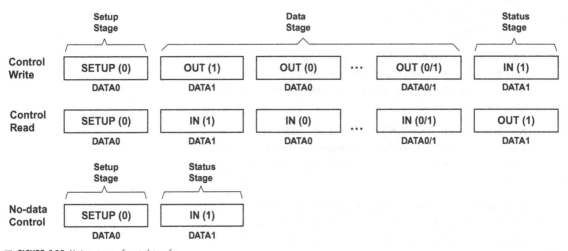

■ **FIGURE 5.35** Various types of control transfers.

Interrupt transfers. The interrupt in USB is a little different from the normal concept of interrupt used in the context of microprocessors; here, if a device requires the attention of the host, it must wait until the host polls it before it can report that it has urgency of transfer. Interrupt transfers provide the following:

- Guaranteed latency
- Stream pipe, unidirectional
- Error detection and next period retry

Interrupt transfers are typically nonperiodic, small-sized device-"initiated" communication requiring a time-bound response. An interrupt request is queued by the device until the host polls the USB device for any data transfer.

Figure 5.36 shows the format of an Interrupt IN and Interrupt OUT transaction.

- *IN*: The host periodically polls the interrupt endpoint. This rate of polling is specified in the descriptor. Each poll will involve the host

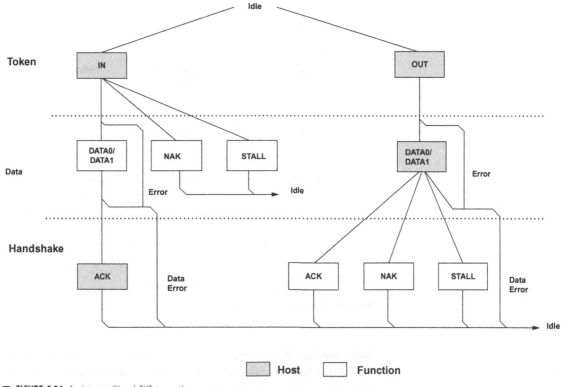

■ **FIGURE 5.36** An interrupt IN and OUT transaction.

sending an IN token. When an interrupt is queued by the device, the function sends a data packet containing data relevant to the interrupt when it receives the IN token. When there is no interrupt to be served, the device responds with a NACK.

■ *OUT*: When the host wants to send the device interrupt data, it issues an OUT token followed by a data packet containing the interrupt data.

Isochronous transfers. Isochronous transfers occur continuously and periodically and have time-sensitive data, such as an audio or video stream. Here, drop in packet due to errors are allowed but not the delay in transaction. Isochronous transfer is unidirectional data transfer with bounded latency. Data being sent on an isochronous endpoint should be less than the prenegotiated size and may vary in length from transaction to transaction.

Figure 5.37 shows the format of an Isochronous IN and OUT transaction. It does not have a handshaking stage or errors stages or STALL/HALT conditions.

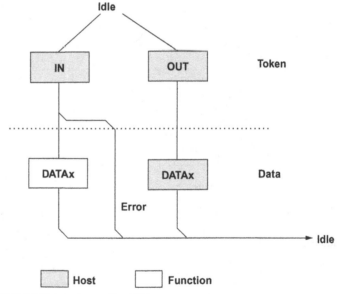

■ **FIGURE 5.37** An isochronous IN and OUT transaction.

Bulk transfers. Bulk transfers are a type of lossless data transfer used for transferring large chunks of data. Bulk transfers provide error correction (CRC16), ensuring data are transmitted and received without error.

Bulk transfers use spare unallocated bandwidth with least priority of allocation. Thus Bulk transfers should only be used for data that is not time sensitive. Bulk transfers are only supported by full- and high-speed devices with maximum bulk packet size of 8, 16, 32, or 64 bytes. For high-speed endpoints, the maximum packet size is 512 bytes long (Figure 5.38).

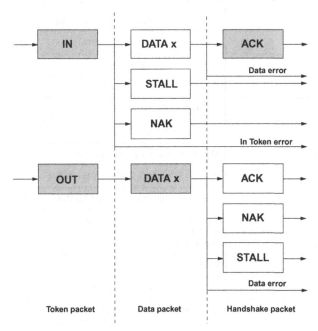

■ **FIGURE 5.38** A bulk IN and OUT transaction.

Figure 5.43 shows the format of a bulk IN and OUT transaction with host initiating IN or OUT Token. The function responds with data. Figure 5.43 indicates all possible packet types during bulk transaction.

Allocation of bandwidth. The host manages the various bandwidths for various types of transfer discussed here during the enumeration process. No more than 90% of any frame can be allocated for periodic transfers (interrupt and isochronous) on a full-speed bus. And 80% (of a micro frame) in a full-speed device, such that the remaining 10% is used for control transfers. The bulk transfers get the remaining bandwidth.

USB descriptors. The device descriptor table inside a device informs the host about its various capabilities. The USB organization has standardized the device descriptor table.

Device descriptor. Various types of information available in device descriptors are shown in Table 5.8 and described in the subsequent paragraphs. Table 5.8 shows the structure for a device descriptor.

Table 5.8 Device Descriptor Table

Offset	Field	Size (Bytes)	Description
0	bLength	1	Length of this descriptor = 18 bytes
1	bDescriptorType	1	Descriptor type = DEVICE (01h)
2	bcdUSB	2	USB Spec Version (BCD)
4	bDeviceClass	1	Device class
5	bDeviceSubClass	1	Device subclass
6	bDeviceProtocol	1	Device Protocol
7	bMaxPacketSize0	1	Max Packet size for endpoint 0
8	idVendor	2	Vendor ID (or VID, assigned by USB-IF)
10	idProduct	2	Product ID (or PID, assigned by the manufacturer)
12	bcdDevice	2	Device release number (BCD)
14	iManufacturer	1	Index of manufacturer string
15	iProduct	1	Index of product string
16	iSerialNumber	1	Index of serial number string
17	bNumConfigurations	1	Number of configurations supported

- *bLength*: Total length in bytes of the device descriptor.
- *bcdUSB*: USB revision that the device supports, which should be latest supported revision. This helps the host to determine which driver to load.
- *bDeviceClass*, *bDeviceSubClass*, and *bDeviceProtocol* are used by the operating system to identify a driver for a USB device during the enumeration process.
- *bMaxPacketSize* reports the maximum number of packets supported by Endpoint zero.
- *iManufacturer*, *iProduct*, and *iSerialNumber* are indexes to string descriptors. String descriptors give information about the manufacturer, product, and serial number.
- *bNumConfigurations* defines the total number of configurations the device can support.

Configuration descriptor. Table 5.9 shows the structure for a configuration descriptor.

Table 5.9 Configuration Descriptor Table

Offset	Field	Description
0	bLength	Length of this descriptor = 9 bytes
1	bDescriptorType	Descriptor type = CONFIGURATION (02h)
2	wTotalLength	Total length including interface and endpoint descriptors (2 bytes)
4	bNumInterfaces	Number of interfaces in this configuration
5	bConfigurationValue	Configuration value used by SET_CONFIGURATION to select this configuration (1 byte)
6	iConfiguration	Index of string that describes this configuration (1 byte)
7	bmAttributes	Bit 7: Reserved (set to 1) Bit 6: Self-powered Bit 5: Remote wakeup
8	bMaxPower	Maximum power required for this configuration (in 2 mA units) (1 Byte)

Interface association descriptor (IAD). This descriptor describes two or more interfaces that are associated with a single device function. The IAD informs the host that the interfaces are linked together. This descriptor is not required in all cases and is useful only when the device has two or more interfaces (see Table 5.10).

Table 5.10 Interface Association Descriptor Table

Offset	Field	Size (Bytes)	Description
0	bLength	1	Descriptor size in bytes
1	bDescriptorType	1	Descriptor type = INTERFACE ASSOCIATION (0Bh)
2	bFirstInterface	1	Number identifying the first interface associated with the function
3	bInterfaceCount	1	The number of contiguous interfaces associated with the function
4	bFunctionClass	1	Class code
5	bFunctionSubClass	1	Subclass code
6	bFunctionProtocol	1	Protocol code
7	iFunction	1	Index of string descriptor for the function

Interface descriptor. An interface descriptor describes a specific interface within a configuration, defines the USB class of the device and its functionality, and helps in identifying proper driver.

Table 5.11 shows the structure for an interface descriptor.

Table 5.11 Interface Descriptor Table

Offset	Field	Description
0	bLength	Length of this descriptor = 9 in this case
1	bDescriptorType	Descriptor type = INTERFACE (04h)
2	bInterfaceNumber	Zero-based index of this interface
3	bAlternateSetting	Alternate setting value
4	bNumEndpoints	Number of endpoints used by this interface (not including EP0)
5	bInterfaceClass	Interface class
6	bInterfaceSubClass	Interface subclass
7	bInterfaceProtocol	Interface protocol
8	iInterface	Index to string describing this interface

Endpoint descriptor. This descriptor gives the endpoint information and includes direction of the endpoint, transfer type, and maximum packet size. Table 5.12 shows the structure for an endpoint descriptor.

Table 5.12 Endpoint Descriptor

Offset	Field	Description
0	bLength	Length of this descriptor = 7 in this case
1	bDescriptorType	Descriptor type = ENDPOINT (05h)
2	bEndpointAddress	Bit 3...0: The endpoint number Bit 6...4: Reserved, reset to zero Bit 7: Direction. Ignored for Control 0 = OUT endpoint 1 = IN endpoint
3	bmAttributes	Bits 1..0: Transfer Type 00 = Control 01 = Isochronous 10 = Bulk 11 = Interrupt If not an isochronous endpoint, bits 5...2 are reserved and must be set to zero. If isochronous, they are defined as follows: Bits 3..2: Synchronization Type 00 = No Synchronization 01 = Asynchronous

(Continued)

Table 5.12 Endpoint Descriptor *(Continued)*

Offset	Field	Description
		10 = Adaptive 11 = Synchronous Bits 5..4: Usage Type 00 = Data endpoint 01 = Feedback endpoint 10 = Implicit feedback Data endpoint 11 = Reserved
4	wMaxPacketSize	Maximum packet size for this endpoint
6	bInterval	Polling interval in milliseconds for interrupt endpoints (1 for isochronous endpoints, ignored for control or bulk)
7	bRefresh	1 Number Reset to 0.
7	bSynchAddress	1 Endpoint Reset to 0.

String descriptor. The string descriptor is an optional descriptor and gives user-readable information about the device. The structure is shown in Table 5.13.

Table 5.13 String Descriptor Table

Offset	Field	Size (Bytes)	Description
0	bLength	1	Length of this descriptor = 7 bytes
1	bDescriptorType	1	Descriptor type = STRING (03h)
2..n	bString or wLangID	Variable	Unicode encoded text string or LANGID code

Other miscellaneous descriptor types

- *Report descriptors*: A USB device class may require an extended set of descriptor information. The descriptor format is present in the class definition specification.
- *MS OS descriptor*: Microsoft has a descriptor called the Microsoft OS Feature Descriptor (also called the MS OS descriptor) and gives Microsoft Windows specific information such as special icons and registry settings. The MS OS descriptor is described on the MSDN website.
- *Device qualifier descriptor*: This describes information required by devices that support 2-speed configurations.

USB class devices

The USB Implementers Forum has a list of approved USB device classes. The most common device classes are Human Interface Device (HID), Mass Storage Device (MSD), Communication Device Class (CDC), and Vendor (Vendor-Specific). Devices that do not meet the definition of a specific USB device class are called vendor-specific devices. These devices allow developers to create customized applications. Table 5.14 shows some USB class codes to give an idea of the various USB classes.

Table 5.14 USB Class Codes

Base Class	Descriptor Usage	Description
00h	Device	Use class information in the interface Descriptors
01h	Interface	Audio
02h	Both	Communications and CDC Control
03h	Interface	HID (Human Interface Device)
05h	Interface	Physical
06h	Interface	Image
07h	Interface	Printer
08h	Interface	Mass Storage
09h	Device	Hub
0Ah	Interface	CDC-Data
0Bh	Interface	Smart Card
0Dh	Interface	Content Security
0Eh	Interface	Video
0Fh	Interface	Personal Healthcare
10h	Interface	Audio/Video Devices
11h	Device	Billboard Device Class
DCh	Both	Diagnostic Device
E0h	Interface	Wireless Controller
EFh	Both	Miscellaneous
FEh	Interface	Application Specific
FFh	Both	Vendor Specific

Summary of USB enumeration and configuration

Let's summarize all that has been discussed in this chapter for the process of enumeration and configuration of a USB device. Enumeration is actually one part in a three-stage process:

- Dynamic detection
- Enumeration
- Configuration

The pull-down resistors on the host/hub side, which alter the voltage in the USB pin, helps in determining whether a device is plugged in and what the supported speed is. This is called *Dynamic detection*. After this, *Enumeration* helps in assigning a unique address to a newly attached device. *Configuration* helps in determining a device's capabilities. The requests that the host uses to learn about a device are called *standard requests* and must support these requests on all USB devices. The host learns of the newly attached device by using an interrupt endpoint to get a report about the hub's status. This includes changes in port status using the GET_PORT_-STATUS request. After this it detects the speed of the device. Initially only full speed or low speed is detected by the hub by detecting if the pull-up resistor is on the D+ or D− line. This information is then reported to the host by another GET_PORT_STATUS request. The host issues a SET_-PORT_FEATURE request to the hub to reset the newly attached device. The device is put into a reset state by pulling both the D+ and D− lines down to GND (0 V) for more than 2.5 µs. This reset state is held for 10 ms by the hub. During this reset, a series of J-State and K-State occurs to determine whether the device supports high speed. For high speed, it issues a single K-State. A high-speed hub detects this K-State and responds "KJKJKJ" pattern. The device detects this pattern and removes its pull-up resistor from its D+ line. The device can now respond to requests from the host in the form of control transfers to its default address of 00h. The host starts by issuing a GET_DESCRIPTOR request to the device default pipe. The USB specification requires that a device return at least 8 bytes of the device descriptor, when requested, if the device has the default address of 00h. The host applies an address to the device with the SET_ADDRESS request. All communication beyond this point will use the new address. The device is now in the address state. The host issues a command, GET_-DESCRIPTOR, using the newly assigned address, to read the descriptors from the device. Next the host issues another GET_DESCRIPTOR command for the configuration descriptor. This request not only returns the configuration descriptor, but all other associated descriptors. After all descriptors are received, the host sets a specific device configuration using the SET_CONFIGURATION request. Devices that support multiple configurations can allow the user or the driver to select the proper configuration. The device is now in the configured state and ready for use in an application.

GRAPHICS COMPONENTS AND THEIR SIGNIFICANCE IN THE SoC

As we already discussed in Chapter 3, by graphics we refer to the system components involved in 3D rendering and media encode/decode/transcode. So the data flow on the computer system will look like Figure 5.39.

In Figure 5.39, the thick arrows represent the data flow while thin arrows represent the control flow. So, essentially what the diagram illustrates is that the CPU provides the data and control for the graphics device. The graphics device picks the data up from memory and works as per controls from the CPU and writes the data back to the memory. The display controller picks the data from memory and works as per control from the CPU and displays the content on to the display panel/device. It must be noted that there is no direct connection or control/data-flow interaction between graphics and display controller. It must also be noted that in the absence of a graphics processor, data are being generated by the CPU. As the need for sophisticated graphics processing kept on growing due to fancy GUI and video gaming usages, the graphics processor continued to become more and more powerful. And since the graphics processors processed huge data, the interfaces used for connecting the graphics device to the system continued to evolve or change to meet the bandwidth requirement of the graphics device.

In the following sections we will go over the graphics interfaces as they evolved.

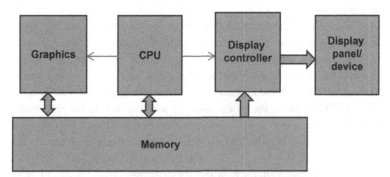

■ **FIGURE 5.39** Control and data flow between CPU, graphics, and display controller.

ISA, EISA, and MCA bus

ISA stands for industry standard architecture. This is the oldest of PC interfaces to hold graphics devices. ISA is also known as the AT bus. This interface is no longer used for graphics cards; in fact, rarely do we now find a motherboard with an ISA slot. ISA came in two flavors: 8-bit and 16-bit. The bus was further extended to 32 bits and given the name EISA or extended ISA bus. EISA also supported "bus mastering" so it could do direct memory access (DMA).

The ISA bus was developed by IBM as an 8-bit bus around 1981, and then it was followed up with 16-bit extension in 1984. The 8-bit version operated at 4.77 MHz while the 16-bit version operated at 6 or 8 MHz. ISA was an open specification. However, to fix the limitations of the ISA bus, IBM developed the MCA (microchannel architecture) bus. MCA was significantly advanced when compared to ISA, but, unlike ISA, it was a closed or proprietary specification. So, in parallel to IBM's proprietary MCA standard, the computing industry responded with EISA as a 32-bit extension of ISA that supported bus mastering. However, only 16 MB of the main memory was accessible through DMA. The EISA bus was fully backward compatible with ISA.

There were significant limitations with respect to using the ISA buses on the system. The system users needed to know the hardware details (like which interrupt line a device was using, I/O address, or DMA channels) of the devices sitting on the ISA bus, and they had to supply that information while configuring the devices to work. Essentially, the ISA bus and devices on them were not plug and play (PnP) capable. MCA did not have the limitation with respect to the PnP behavior.

VESA local bus

In an attempt to solve the bandwidth limitation for graphics, VESA (Video Electronics Standards Association) proposed the VESA Local Bus (VLBus). The VLBus strategy is to connect the video controller and any other high-bandwidth device to the system bus directly. Through a buffer, up to three devices could be connected. A typical VLBus system architecture is shown in Figure 5.40.

VLBus temporarily solved the problem of bandwidth; however, since it was connected to the system bus directly, it was processor specific. VESA also tried to resolve the configuration problem that existed with ISA by mandating that VLBus devices be autoconfigurable. Again, however, there was no standard for autoconfiguration, and therefore each VLBus device manufacturer or vendor defined their own mechanism of autoconfiguration. In addition to that,

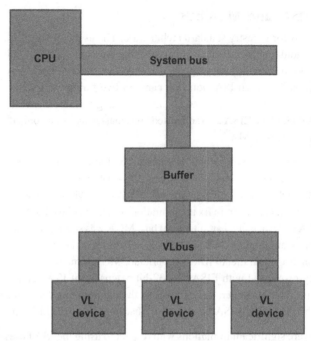

■ **FIGURE 5.40** VLBus architecture.

VESA did not specifically define the electrical characteristics of the VLBus devices. The devices were just assumed to be compatible with the system bus of the time. All put together, the VLBus was tied to the processor bus of the time (486) and went out of the market with the advent of Pentium.

Peripheral Components Interconnect (PCI)

As discussed in the previous sections, there were significant limitations to ISA and VLBus buses, and PCI was the next attempt to solve the problem. Intel developed the PCI standard and the first version came out in 1992. There were two fundamental goals for development of PCI: fix the limitations of existing buses, primarily the bandwidth and autoconfigurability. In addition, there was a need to get rid of the dependence on the system bus so the devices did not have to change as the processor/system bus did change.

PCI stands for Peripheral Components Interconnect. It is a 32-bit-wide bus that runs at 33 MHz and delivers a throughput of 1.33 MB/s. PCI is the standard for most computer AICs (add-in cards) today; however, the bandwidth of PCI is limiting for modern graphics cards. Therefore newer buses were developed to enhance the bandwidth while keeping the backward compatibility intact. From the connection standpoint the system architecture of a PCI-based system looks like Figure 5.41. It must be noted that the diagram

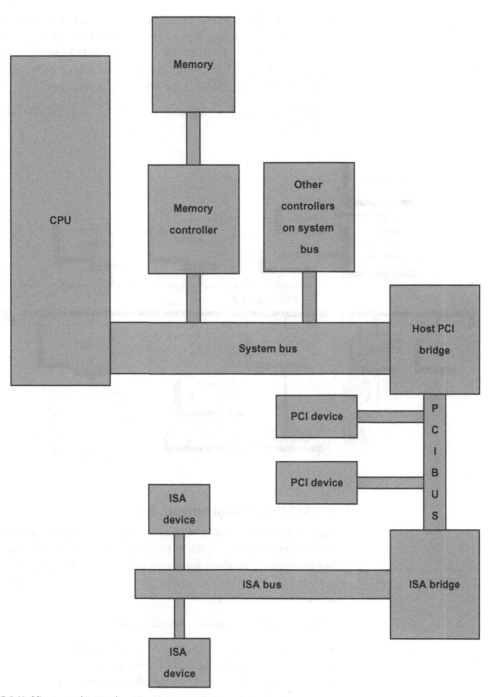

■ **FIGURE 5.41** PCI system architecture diagram.

is only to illustrate the PCI connection and avoids lots of other details in terms of connections and devices. As shown in Figure 5.41, the Host PCI bridge takes care of converting from system bus to PCI. And all PCI devices are hooked on the PCI bus, which is on the other side of the bridge. For any change in the system bus only the bridge needs to change.

A slightly more detailed diagram from the PCI 2.3 specification is shown in Figure 5.42.

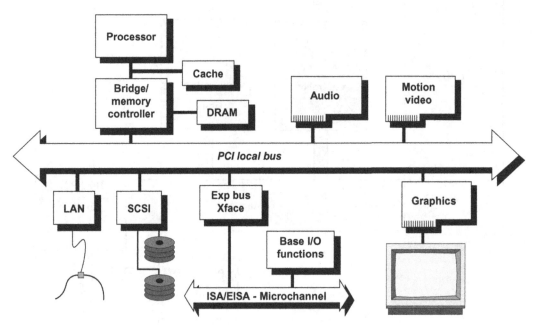

■ **FIGURE 5.42** PCI system.

PCI-SIG (PCI Special Interest Group)

Though the first version of PCI was developed by Intel, the specification is revised and maintained by the PCI Special Interest Group (PCI-SIG), which has hundreds of members worldwide representing all aspects of the computer industry: chip vendors, motherboard suppliers, BIOS and OS vendors, add-in card suppliers, and so on.

PCI is covered by a gamut of specifications:

■ PCI local bus specification (revision 3.0 is current)
■ Mobile design guide (revision 1.1 is current)
■ Power management interface specification (revision 1.1 is current)
■ PCI to PCI bridge architecture specification (revision 1.1 is current)

- PCI hot-plug specification (revision 1.0 is current)
- Small PCI specification (revision 1.5a is current)
- PCI BIOS specification (revision 2.1 is current)
- PCI-X protocol addendum to PCI specification (revision 2.0 is current)
- PCI Express specification (revision 1.0 is current)

PCI features

Next we will discuss the design principles of PCI interface.

High performance. Variations of PCI specification support from 132 to 1064 MB/s. As of specification 2.3, the bandwidth is as shown in Table 5.15.

Table 5.15 Throughput of Different Versions of PCI Buses	
32 Bit at 33 MHz	132 MB/s
64 Bit at 33 MHz	264 MB/s
32 Bit at 66 MHz	264 MB/s
64 Bit at 66 MHz	532 MB/s
32 Bit at 133 MHz	532 MB/s
64 Bit at 133 MHz	1064 MB/s

In addition to this, PCI also employs various other measures to improve performance:

- Variable-length linear and cacheable wrap mode bursting for both read and writes improve write-dependent graphics performance.
- Low-latency random access (60-ns write access latency for 33 MHz PCI to 30 ns for 133-MHz PCI-X to slave registers from master parked on bus).
- Capable of full concurrency with processor/memory subsystem.
- Synchronous bus with operation up to 33, 66, or 133 MHz.
- Hidden central arbitration.

Low cost. PCI has been optimized for direct silicon interconnection, meaning that there is no glue logic required. Electrical and frequency specification meets the standard ASIC technologies and other typical processes. In addition, the pins are multiplexed and therefore reduce the total number of pins required.

Ease of use. PCI devices contain information in registers that is required for configuration of the device. This methodology enables for autoconfiguration of devices—one of the biggest challenges with ISA devices.

Longevity. As discussed earlier, the bridge between system bus and PCI bus makes PCI processor independent. It can be, and in fact has been, used across multiple families of processors and it will be for future generations. The 64-bit support allows it to work with 64-bit processor architectures seamlessly. Also, this works on both 5 and 3.3 V, so there is no threat to PCI due to voltage levels moving from 5 to 3.3 V.

Interoperability and reliability. There are multiple features that support the interoperability and reliability. Small form factor add-in cards, along with forward and backward interoperability between various combinations of data-width (32 and 64-bit) and frequency (33, 66, and 132 MHz) combinations make it interoperable. And the reliability aspect is covered by extensive hardware model validation. In addition, the signals in the PCI specification allow for power supplies to be optimized for the expected system usage by monitoring add-in cards that could surpass the maximum power budgeted by the system.

Flexibility. Full multimaster capability allows peer-to-peer transactions.

Data integrity. In the PCI specification, parity logic is supported for both data and address lines, so errors on data or address buses can be detected. This allows for robust client platform implementation.

Software compatibility. Various revisions of PCI standards are software transparent. That means that the software written for a device using one revision of PCI can be used as is for the device when it moves to use a newer revision of the PCI bus specification.

PCI signal definition

To handle data, addressing, interface control, arbitration, and system functions, PCI requires a minimum of 47 pins for a target-only device and 49 pins for devices that want to be bus master. Figure 5.43 shows the pinout of a PCI-compliant device. The diagram shows the required pins on one side, and optional pins on the other side.

It must be noted that active low signals are marked with "#" at the end: for example, RST#. Based on the functionalities, these PCI signals can be grouped as follows:

System. CLK provides the clock or timing for any and every transaction and is an input to the device. All the signals except for RST#, interrupt (INTA#, INTB#, INTC#, and INTD#), CLKRUN#, PME# are sampled at the rising edge of CLK.

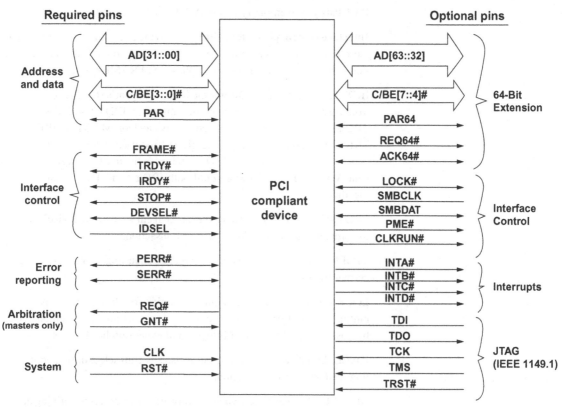

■ **FIGURE 5.43** PCI Pin list.

RST# Reset is used to bring PCI-specific registers, sequencers, and signals
to a consistent state. Any time RST# is asserted, all PCI output signals must
be driven to their benign state.

Address and data pins. AD[31::0] address and data are multiplexed on the
same PCI pins. A bus transaction consists of an address phase followed by
one or more data phases. PCI supports both read and write burst cycles.
Burst cycles mean that the device can transmit data repeatedly without going
through all the steps required to transmit each piece of data in a separate
transaction. The usual reason for having a burst mode capability, or using
burst mode, is to increase data throughput. The address phase is the first
clock cycle in which FRAME# is asserted.

C/BE[3::0]# Bus commands and "byte enables" are multiplexed on the same
PCI pins. During the address phase of the transaction C/BE[3::0]# defines the
bus command while during data phase the same is used for "Byte Enable."

PAR Parity is even parity across AD[31::0] and C/BE[3::0]#.

Interface control pins. FRAME# Cycle Frame is driven by the current master to indicate a transaction. FRAME# is asserted to indicate a bus transaction is beginning. In the final data phase the FRAME# is de-asserted.

IRDY#: Initiator ready indicates the initiating agent's (bus master's) ability to complete the current data phase of transaction. IRDY# is used along with TRDY#. A data phase is completed on any clock on which both IRDY# and TRDY# are asserted. During write, IRDY# indicates valid data is present on AD[31::0]. During read, the IRDY# indicates the master is ready to accept data. Wait cycles are inserted until both IRDY# and TRDY# are asserted together.

TRDY#: Target ready indicates the target agent's (selected device's) ability to complete the current data phase of transaction.

STOP#: Stop indicates that current target is requesting the master to stop the current transaction.

LOCK#: Lock indicates an atomic operation to a bridge that may require multiple transactions to complete. Locked transactions can be initiated only by host bridges, PCI-to-PCI bridges, and expansion bus bridges.

IDSEL: Initialization device select is used as chip select during configuration read and write transaction.

DEVSEL#: Device select, when actively driven, indicates the driving device has decoded its address as the target of current access. As an input, DEVSEL# indicates whether any device on the bus has been selected.

Arbitration pins. Since PCI is multimaster bus, there is need for arbitration before one of the many masters can take control. Arbitration control pins are applicable for bus masters only.

REQ# Request indicates to the arbiter that this agent desires to use the bus. GNT# Grant indicates to the agent that access to the bus has been granted.

Error-reporting pins. PERR# Parity Error is for reporting of data parity errors during all PCI transactions except special cycle.

SERR# System Error is for reporting address parity errors, data parity errors on the special cycle, or any other system error when the result will be catastrophic.

Interrupt pins (optional). There are four pins for the PCI device to raise interrupt and seek attention. There are four pins as, by definition, PCI device can be multifunctional and therefore can allow different functions to use different pins for interrupt. The "Interrupt Pin" register in PCI configuration space defines which INTx line the function uses to raise interrupt:

- INTA# interrupt A is used to raise interrupt
- INTB# interrupt B is used to raise interrupt
- INTC# interrupt C is used to raise interrupt
- INTD# interrupt D is used to raise interrupt

Additional signals. PRSNT# Present signals are not signals for a device but are provided by an add-in card. This signal is used to indicate the system board whether or not an add-in card is physically present in the slot and if present what the total power requirement of the add-in card is.

CLKRUN# Clock running is an optional signal, used as an input for a device to determine the status of CLK.

M66EN: This pin indicates to the device whether the bus segment is operating at 66 or 33 MHz.

PME# The Power Management Event signal is an optional signal that can be used by a device to request a change in device or system power state.

3.3Vaux: An optional 3.3-V auxiliary power source that delivers power to the PCI add-in card for generation of PME# signal when the main power to the card has been turned off by software.

64-bit bus extension signals (optional). Additional signals similar to the ones for 32-bit support are required for 64-bit extension. The signals are AD[63::32], C/BE[7::4]#, REQ64#, ACK64#, and PAR64.

JTAG/boundary scan pins (optional). Signals to support IEEE 1149.1, test access port, and boundary scan architecture are included as an optional interface for PCI devices. The signals are TCK, TDI, TDO, TMS, and TRST#, and they have the same meaning as the IEEE 1149.1 standard defines.

SMBus interface pins (optional). SMBCLK and SMBDAT signals are included as optional pins to support SMBus.

Sideband signals (optional). PCI allows for sideband communication signals for product-specific functional/performance enhancement specifications. A sideband signal is loosely defined as any signal not part of the PCI specification that connects two or more PCI-compliant agents and has meaning only to these agents. Sideband signals are permitted for two or more devices to communicate some aspect of their device-specific state in order to improve the overall effectiveness of PCI utilization or system operation. No pins are allowed in the PCI connector for sideband signals. Therefore, sideband signals must be limited to the system board environment.

Furthermore, sideband signals may never violate the specified protocol on defined PCI signals or cause the specified protocol to be violated.

PCI bus operation

Commands. Fundamental to PCI bus operation are PCI commands. PCI commands are sent over C/BE[3::0]#, and the encoding is defined as shown in Table 5.16. These commands are discussed briefly in the following section. Additional details, however, can be found in PCI specification.

Table 5.16 PCI Command Encoding

C/BE[3::0]#	Command Type
0000	Interrupt Acknowledge
0001	Special Cycle
0010	I/O Read
0011	I/O Write
0100	Reserved
0101	Reserved
0110	Memory Read
0111	Memory Write
1000	Reserved
1001	Reserved
1010	Configuration Read
1011	Configuration Write
1100	Memory Read multiple
1101	Dual Address Cycle
1110	Memory Read Line
1111	Memory Write and Invalidate

The Interrupt Acknowledge command is used to acknowledge an interrupt to the system interrupt controller. This is done via a read that is implicitly addressed to the system interrupt controller.

The Special Cycle command is a simple message broadcast mechanism on PCI. The purpose of special cycle is to provide a mechanism for sideband communication without having a physical signal dedicated for sideband communication.

The I/O Read and I/O Write commands are used to read and write data (respectively) from a device that is mapped in I/O Address Space. For this command, the address of read/write is specified by AD[31::00].

Similar to I/O Read/Write, the Memory Read/Write command is used to read/write data from a device that is mapped in the Memory Address Space. Similarly, for this command, the address of read/write is specified by AD [31::00].

The Configuration Read/Write command is used to read/write the Configuration Space of a PCI device. The device for configuration access is selected by asserting its IDSEL signal. At the time of configuration space access, the AD [1::0] are 00.

The Memory Read Multiple command is to improve the performance for bulk transfers. It is fundamentally similar to the Memory Read command; however, it additionally indicates that the master may fetch more than one cache line before disconnecting. The memory controller continues to pipeline memory requests as long as FRAME# is asserted. This command can be used for prefetching data to improve performance by avoiding the latency.

The Memory Read Line command is fundamentally similar to the Memory Read command; however, it additionally indicates that the master intends to fetch a complete cache line. This command again is used to improve performance.

The Memory Write and Invalidate command is fundamentally similar to the Memory Write command; however, it guarantees that a minimum of one cacheline transfer will be completed.

The Dual Address Cycle command is used to support 64-bit address to devices that support 64-bit addressing.

Commands marked as Reserved are not used currently and are available for future use.

PCI protocol fundamental. The basic bus transfer mechanism on PCI is a burst. A burst cycle is composed of an address phase, which is followed by one or more data phases. Bursts are supported by PCI for both Memory and I/O Address Spaces.

Basic transfer control. All PCI data transfers are controlled with three signals: FRAME#, IRDY#, TRDY#. Based on the combination of the states of these signals, data transaction are initiated and terminated. For example, Figure 5.44 shows a read transaction and starts with an address phase that occurs when FRAME# is asserted for the first time and occurs on clock 2. During the address phase, AD[31::00] contain a valid address and C/BE

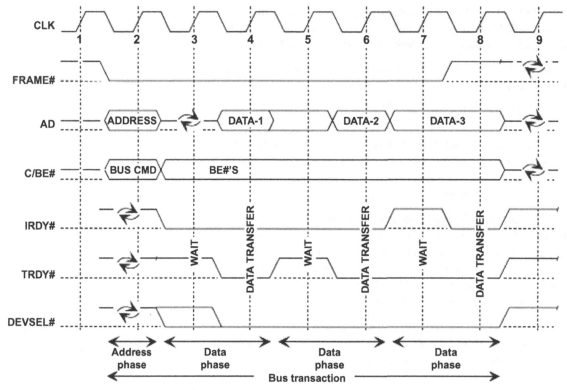

■ **FIGURE 5.44** Basic Read transaction.

[3::0]# contain a valid bus command. The first clock of the first data phase is clock 3. During the data phase, C/BE# indicate which byte lanes are involved in the current data phase. A data phase may consist of wait cycles and a data transfer. The C/BE# output buffers must remain enabled (for both read and writes) from the first clock of the data phase through the end of the transaction. This ensures C/BE# are not left floating for long intervals. To illustrate the sequence, Figure 5.44 shows the signals during a Basic Read transaction.

PCI arbitration. Since PCI is a multimaster bus protocol, there has to be a mechanism of deciding who gets the control of the bus when many of them are contesting.

The process is simple; any agent wishing to initiate a PCI transaction has to request the control and be granted to control the bus. There is a pair of signals REQ# and GNT# for the same. Only one GNT# signal can be asserted at any instant of time. The agent who sees its GNT# asserted may initiate a bus transaction when it detects that the bus is idle. The bus idle state is denoted by FRMAE# and IRDY# both being de-asserted.

A master device is only allowed to assert its REQ# when it actually needs the bus to execute a transaction. In other words, it is not allowed to continuously assert REQ# in order to monopolize the bus. The arbiter may be designed to "park" the bus on a default master when the bus is idle. This is accomplished by asserting GNT# to the default master when the bus is idle. The agent on whom the bus is parked can initiate a transaction without first asserting REQ#. This saves one clock.

In order to ensure fairness in bus usage across multiple agents, there are PCI Latency Timers that are the mechanism for PCI Bus-Mastering devices to share the PCI bus fairly. "Fair" in this case means that devices will not use such a large portion of the available PCI bus bandwidth, that other devices are not able to get needed work done. The way this works is that each PCI device that can operate in bus-master mode is required to implement a timer, called the Latency Timer, that limits the time that device can hold the PCI bus. The timer starts when the device gains bus ownership, and counts down at the rate of the PCI clock. When the counter reaches zero, the device is required to release the bus. If no other devices are waiting for bus ownership, it may simply grab the bus again and transfer more data.

PCI bridges. Because of electrical loading issues, the number of devices that can be supported on a given bus segment is limited. To allow systems to be built beyond a single bus segment, PCI-to-PCI bridges are defined. A PCI-to-PCI bridge requires a mechanism to know how and when to forward configuration accesses to devices that reside behind the bridge. Figure 5.45 illustrates an example system architecture with PCI-2-PCI bridge.

Addressing. PCI defines three physical address spaces. The memory and I/O address spaces for any devices are well known. PCI defines configuration address space to support the hardware configuration.

PCI devices are required to implement BAR (Base Address Registers) to request a range of addresses, which can be used to provide access to internal registers or functions. The configuration software like BIOS uses the BAR register to determine how much space a device requires in a given address space and then assigns where in that space the device will reside. When a transaction is initiated on the interface, each potential target compares the address with its Base Address register(s) to determine whether it is the target of the current transaction. If it is the target, the device asserts DEVSEL# to claim the access.

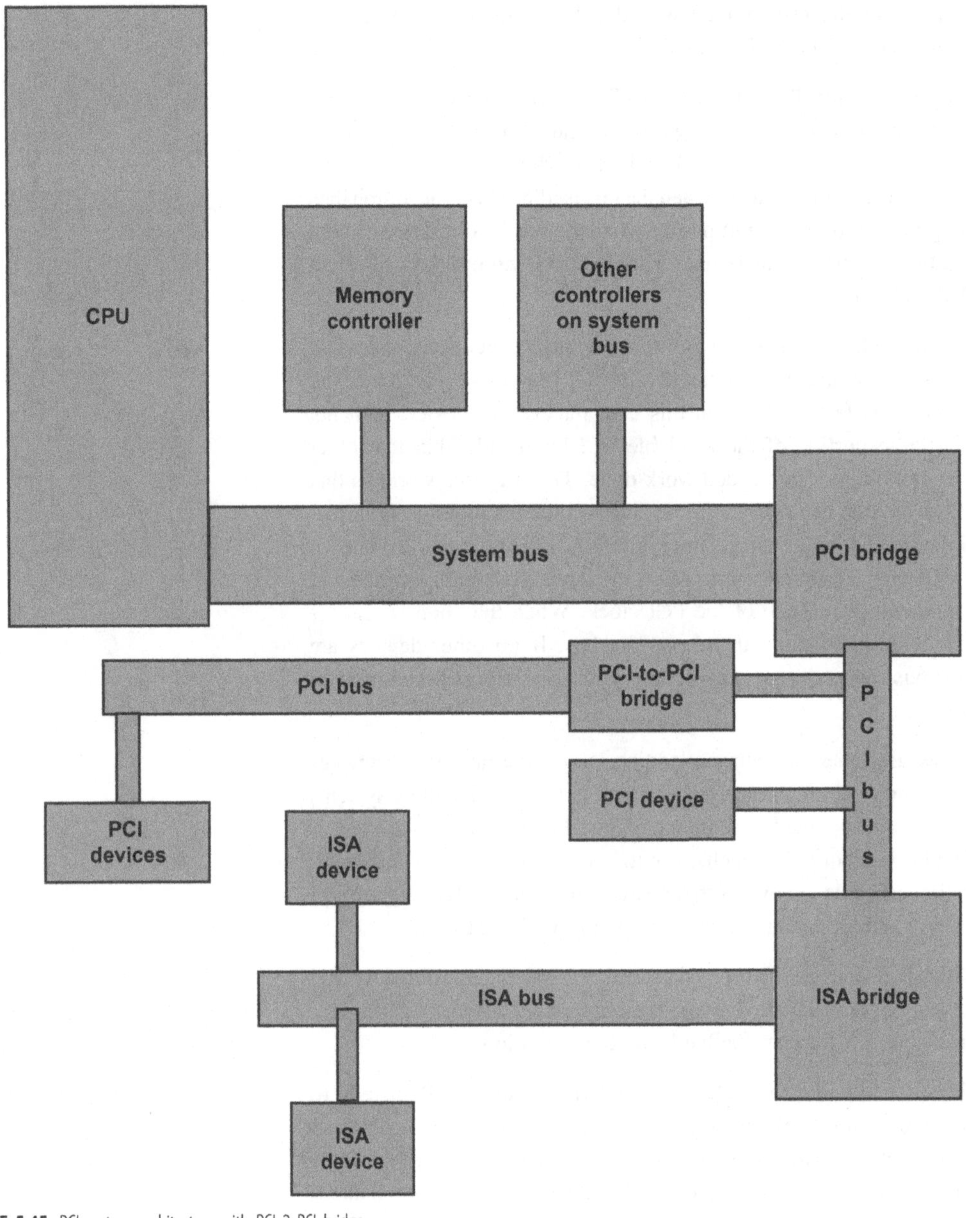

■ **FIGURE 5.45** PCI system architecture with PCI-2-PCI bridge.

IO space decoding. In the I/O Address Space, all 32 AD lines are used to provide a full byte address. The master that initiates an I/O transaction is required to ensure that AD[1::0] indicate the least significant valid byte for the transaction.

The byte enables indicate the size of the transfer and the affected bytes within the DWORD and must be consistent with AD[1::0].

Memory space decoding. In the Memory Address Space, the AD[31::02] bus provides a DWORD aligned address. AD[1::0] are not part of the address decode. However, AD[1::0] indicate the order in which the master is requesting the data to be transferred. Table 5.17 lists what the various AD [1:0] combinations mean.

Table 5.17 List of Modes for Memory Access Requests

AD1	AD0	Burst Order
0	0	Linear incrementing
0	1	Reserved
1	0	Cacheline wrap mode
1	1	Reserved

All targets are required to check AD[1::0] during a memory command transaction and either provide the requested burst order or terminate the transaction with Disconnect in one of two ways. The target can use Disconnect with Data during the initial data phase or Disconnect without Data for the second data phase. With either termination, only a single data phase transfers data. The target cannot terminate the transaction with Retry just because it does not support the burst order requested. If the target does not support the burst order requested by the master, the target must complete one data phase and then terminate the request with Disconnect. This ensures that the transaction will complete; however, it may be slow since each request will complete as a single data phase transaction and not as a burst transfer. If a target supports bursting on the bus, the linear burst ordering is a must to support. However, support for "cacheline wrap" mode is optional.

Configuration space decoding. Every device, other than host bus bridges, must implement Configuration Address Space. Host bus bridges may optionally implement Configuration Address Space. In the Configuration Address Space, each function is assigned a unique 256-byte space that is accessed differently than I/O or Memory Address Spaces. The configuration space contains the data to indicate basic information for device: for example, manufacture, resources, capability, and routing information.

Configuration commands (Type 0 and Type 1). To support hierarchical PCI buses, two types of configuration transactions are used. They have the formats illustrated in Figure 5.46 showing the interpretation of AD lines during the address phase of a configuration transaction.

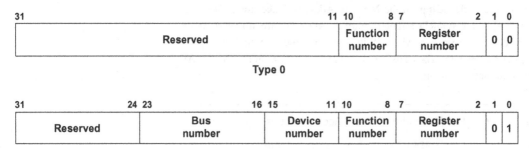

FIGURE 5.46 Address phase formats of configuration transactions.

Type 1 and Type 0 configuration transactions are differentiated by the values on AD[1::0]. A Type 0 configuration transaction (when AD [1::0] = "00") is used to select a device on the bus where the transaction is being run. A Type 1 configuration transaction (when AD[1::0] = "01") is used to pass a configuration request to another bus segment.

Bridges (both host and PCI-to-PCI) that need to generate a Type 0 configuration transaction use the Device Number to select which IDSEL to assert. The Function Number is provided on AD[10::08]. The Register Number is provided on AD[7::2]. AD[1::0] must be "00" for a Type 0 configuration transaction. A Type 0 configuration transaction is not propagated beyond the local PCI bus and must be claimed by a local device or terminated with Master-Abort.

If the target of a configuration transaction resides on another bus (not the local bus), a Type 1 configuration transaction must be used. All targets except PCI-to-PCI bridges ignore Type 1 configuration transactions. PCI-to-PCI bridges decode the Bus Number field to determine if the destination bus of the configuration transaction resides behind the bridge. If the Bus Number is not for a bus behind the bridge, the transaction is ignored. The bridge claims the transaction if the transaction is to a bus behind the bridge. If the Bus Number is not to the secondary bus of the bridge, the transaction is simply passed through unchanged. If the Bus Number matches the secondary bus number, the bridge converts the transaction into a Type0 configuration transaction. The bridge changes AD[1::0] to "00" and passes AD[10::02] through unchanged. The Device Number is decoded to select one of 32 devices on the local bus. The bridge asserts the correct IDSEL and initiates a Type 0 configuration transaction.

Software generation of configuration transactions. Systems must provide a mechanism that allows software to generate PCI configuration transactions. This mechanism is typically located in the host bridge. Two DWORD I/O locations are used to generate configuration transactions for PC-AT-compatible systems. The first DWORD location (CF8h) references a read/write register that is named CONFIG_ADDRESS. The second DWORD address (CFCh) references a read/write register named CONFIG_DATA. The CONFIG_ADDRESS register is 32 bits with the format shown in Figure 5.47.

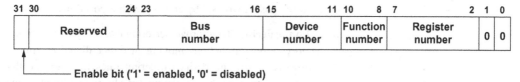

■ **FIGURE 5.47** Layout of CONFIG_ADDRESS register.

Bit 31 is an enable flag for determining when accesses to CONFIG_DATA are to be translated to configuration transactions on the PCI bus. Bits 30 to 24 are reserved, read-only, and must return 0s when read. Bits 23 through 16 choose a specific PCI bus in the system. Bits 15 through 11 choose a specific device on the bus. Bits 10 through 8 choose a specific function in a device (if the device supports multiple functions). Bits 7 through 2 choose a DWORD in the device's Configuration Space. Bits 1 and 0 are read-only and must return 0s when read.

When the host bridge sees a full DWORD I/O write from the host to CONFIG_ADDRESS, the bridge latches the data into its CONFIG_ADDRESS register. On full DWORD I/O reads to CONFIG_ADDRESS, the bridge returns the data in CONFIG_ADDRESS. However, when a host bridge sees an I/O access falling inside the DWORD beginning at CONFIG_DATA address, it checks the Enable bit and the Bus Number in the CONFIG_ADDRESS register. If the Enable bit is set and the Bus Number matches the bridge's Bus Number or any Bus Number behind the bridge, a configuration cycle translation is done.

Accesses in the Configuration Address Space require device selection decoding to be done externally, and based on that the right device has to be signaled via initialization device select, or IDSEL. The IDSEL functions as a classical "chip select" signal. Each device has its own IDSEL input. The exceptions are host bus bridges; they can implement their initialization device selection internally.

Performance improvement tricks

PCI defines and implements a few tricks to improve the overall system performance. We will discuss these in the following section.

Posted writes. Generally, when a bus bridge sees a transaction on one bus that must be forwarded to the other, the original transaction must wait until the forwarded transaction completes before a result is ready. However, to improve performance, in the case of memory writes, the bridge may record the write data internally and signal completion of the write before the forwarded write has completed. Such "sent but not yet arrived" writes are referred to as "posted writes," by analogy with a postal mail message.

Transaction ordering. To improve performance, bridges are permitted to post memory write transactions moving in either direction through the bridge. However, there are defined ordering rules to guarantee that the results of one master's write transactions are observable by other masters in the right (defined) order, even though the write transaction may be posted in a bridge.

Combining, merging, and collapsing. To improve the efficiency of the PCI bus, the PCI standard permits bus bridges to convert multiple bus transactions into one larger transaction under certain conditions. There are three different kinds of conversion, which are named as per their behavior: combining, merging, and collapsing.

Combining. Write transactions to consecutive addresses may be combined into a longer burst write, as long as the order of the accesses in the burst is the same as the order of the original writes. It is permissible to insert extra data phases with all byte enables turned off if the writes are almost consecutive.

Merging. Multiple writes to disjoint portions of the same DWORD may be merged into a single write with multiple byte enables asserted.

Collapsing. Collapsing happens when a sequence of memory writes to the same location (byte, word, or DWORD address) are collapsed into a single bus transaction. Collapsing is not permitted by PCI bridges (host, PCI-to-PCI, or standard expansion) except for a few cases.

Delayed transaction. Delayed Transaction termination is used by targets that cannot complete the initial data phase within the requirements of the

PCI specification. In general, I/O controllers will handle only a single Delayed Transaction at a time, while bridges may choose to handle multiple transactions to improve system performance.

A Delayed Transaction progresses to completion in three steps:

1. Request by the master
2. Completion of the request by the target
3. Completion of the transaction by the master

During the first step, the master generates a transaction on the bus, the target decodes the access, latches the information required to complete the access, and terminates the request with Retry. During the second step, the target independently completes the request on the destination bus using the latched information from the Delayed Request. The target stores the Delayed Completion until the master repeats the initial request. During the third step, the master rearbitrates for the bus and reissues the original request. The target decodes the request and gives the master the completion status and data if applicable. At this point, the Delayed Completion is retired and the transaction has completed.

The key advantage of a Delayed Transaction is that the bus is not held in wait states while completing an access to a slow device. While the originating master rearbitrates for the bus, in between, other bus masters can use the bus bandwidth that would otherwise be wasted holding the master in wait states. The other advantage is that posted writes are not required to be flushed before the request is accepted.

PCI versus PCI-X

PCI-X, which stands for Peripheral Component Interconnect eXtended, is a fully backward-compatible extension to the PCI standard. It enhances the 32-bit PCI Local Bus for higher bandwidth demanded by servers. It is a double-wide version of PCI, running at up to four times the clock speed, but is otherwise similar in electrical implementation and uses the same protocol.

Accelerated Graphics Port (AGP)

AGP stands for Accelerated Graphics Port. This is a high-bandwidth point-to-point channel designed specifically for graphics cards. It was based on PCI 2.1 specifications; however, contrary to PCI, which is shared bus, AGP is dedicated to a device. The base PCI 2.1 specification was not changed in any way. Instead a few performance-oriented extensions were defined and implemented. These extensions are

- Deeply pipelined memory read and writes operations; this helps hide memory access latency completely.
- Demultiplexing of address and data on the bus; this allows for almost 100% bus efficiency.
- AC timing for 133-MHz data transfer rates; this allows for real data throughput in excess of 500 MB/s.

These enhancements are implemented using "sideband" signals. The PCI Specification has not been modified in any way. The AGP interface specification has been conscious in avoiding the use of any of the "reserved" fields in the PCI Specification. This is to ensure that if a later revision of PCI specification starts to use the reserved fields, AGP should not break. The intent is to utilize the PCI design base as it is and add a few graphics-oriented performance enhancements by making the right tradeoff.

AGP neither replaces nor diminishes the necessity of PCI in the system. AGP is actually completely independent of the PCI bus; it is just that the AGP specification is derived from PCI base. It is an additional connection point in the system, as shown in Figure 5.48. It is intended for the exclusive use of visual display devices; all other I/O devices will remain on the PCI bus. The AGP add-in slot uses a different connector body that is not compatible with the PCI connector; therefore, the PCI and AGP boards are not interchangeable.

The AGP interface specification was developed by Intel independently, without any collaboration with PCI Special Interest Group. It has been neither reviewed nor endorsed by PCI-SIG. In fact, except for the fact that the AGP specification was derived from PCI, there is no connection between the two.

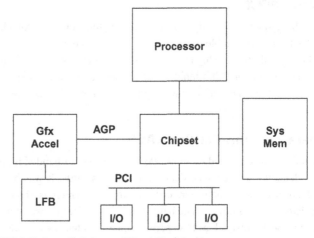

■ **FIGURE 5.48** System Block diagram: AGP and PCI relationship.

AGP operation overview

Memory access pipelining is the key enhancement over PCI protocol provided by the Accelerated Graphics Port (A.G.P. or AGP) Interface Specification. AGP-pipelined bus transactions share most of the PCI signal set; they are interleaved with PCI transactions on the bus. Only memory read and write bus operations targeted at main memory can be pipelined; all other bus operations are executed as PCI transactions. AGP-pipelined operation allows for a single AGP-compliant target, which must always be the system memory controller, which is also referred to as *core logic*. In addition to AGP-compliant target functions, the core logic must also implement a complete PCI sequencer, both master and target. For electrical signaling reasons the AGP is defined as a point-to-point connection. So, there is also only a single AGP-compliant master. The master has to implement AGP-compliant master functions. In addition to that it must also provide full PCI-compliant target functionality. It is optional to implement PCI-compliant master functionality.

Pipeline operation

The AGP interface uses a few newly defined "sideband" control signals, which are used in conjunction with the PCI signal set. AGP-defined protocols are overlaid on the PCI bus at a time and in such a way that a regular non-AGP PCI bus agent would think that the bus is idle. Both pipelined access requests and resulting data transfers are done in this manner. The AGP interface uses both PCI bus transactions as is and the AGP-pipelined transactions. Both of these are interleaved on the same physical connection. The access request in an AGP transaction is signaled differently than in a PCI address phase. The AD and C/BE# signals of the bus are used in both cases, but in the case of AGP it is identified or framed with a new control signal, PIPE#. It is similar to PCI address phases being identified with FRAME#.

In a pipelined operation, the "pipe depth" is defined by number of stages in a pipelined operation. The bus master can effectively maintain the pipe depth by inserting new requests between data replies. The notion of intervening in a pipelined transfer enables the bus master to do that. This notion is similar to running multiple threads on the same physical processor in a time-sliced manner: The context of a thread is stored (backed up) when the processor switches to a different thread and the context of the new thread is loaded. A sample bus sequencing is illustrated in Figure 5.49.

To understand the seamless transition and intermixing between PCI and AGP transactions we need to understand the state transition on the PCI protocol as shown in Figure 5.50.

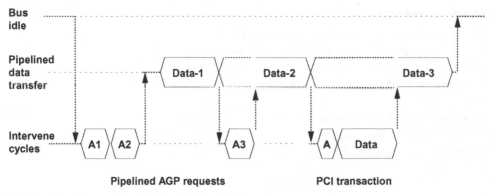

■ **FIGURE 5.49** Illustration of bus sequencing during intervened PCI and AGP transactions.

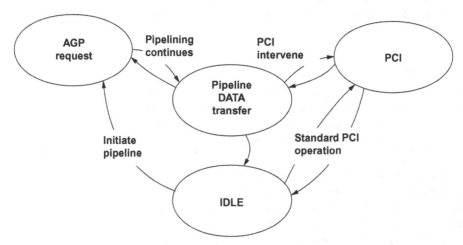

■ **FIGURE 5.50** State transition on the bus.

As can be seen in Figure 5.50, the AGP pipeline is initiated from the idle state after arbitrating for the bus and then issuing AGP access requests (*AGP* state). These requests are transmitted similar to the PCI address phase. The key difference is that they are timed with PIPE# rather than FRAME#. Since the AGP transactions are initiated with PIPE#, regular PCI clients (non-AGP clients) do not recognize the transaction.

Addressing modes and bus operation

Despite the fact that the AGP specification was derived from PCI, there are significant differences between PCI and AGP transactions.

1. The data transfer in AGP transactions is "out of order" "pipelined," and therefore there may be other AGP transactions between the request and associated data transfer. This is in contrast to PCI where a PCI data phase is connected to its associated address phase. And there cannot be any transaction intervening in between. The separation in case of AGP allows the pipe depth to be maintained. It must be noted that all of the access-ordering rules on AGP are based on the arrival order of the access requests, and not the order of actual data transfer.

2. AGP transactions use a completely different set of bus commands (defined as follows) than do PCI transactions. AGP bus commands provide for access *only* to main system memory. PCI bus commands provide for access to multiple address spaces: memory, I/O, configuration. The address space used by AGP commands is the same 32-bit, linear physical space also used by PCI memory space commands, as well as on the processor bus.

3. Memory addresses used in AGP transactions are always aligned on 8-byte boundaries. Also, 8 bytes is the minimum access size, and the length of all accesses are integer multiples of 8 bytes. In contrast, memory accesses for PCI transactions have 4-byte granularity, and they are aligned on 4-byte boundaries. Smaller or odd-size reads in AGP must be accomplished with a PCI read transaction. Smaller or odd-size writes can be accomplished via the C/BE# signals. C/BE# signal acts as a mask, which enables the actual writing of individual bytes within an 8-byte field.

4. AGP accesses do not guarantee memory coherency; that is, AGP accesses do not have to be snooped. AGP access requests have an explicitly defined access length or size whereas the PCI transfer lengths are not explicitly defined and rather are defined by the duration of FRAME#.

5. PCI memory accesses always ensure a coherent view of memory and must be used on accesses where coherency is required.

The format of a complete AGP bus request is shown in Figure 5.51.

The "LLL" field contains the access length in units of Q-words (8 bytes), and displaces the low-order 3 bits of address. So, for example, a length field

■ **FIGURE 5.51** Layout of an AGP access request.

of "000" means that a single Q-word data is being requested, whereas a length field of "111" would indicate 8 Q-words of data being requested. The "CCCC" field contains the bus operation or command as itemized in Table 5.18.

Table 5.18 AGP Bus Commands

CCCC	AGP Operation
0000	Read
0001	Read (High Priority)
0010	Reserved
0011	Reserved
0100	Write
0101	Write (High Priority)
0110	Reserved
0111	Reserved
1000	Long Read
1001	Long Read (High Priority)
1010	Flush
1011	Reserved
1100	Fence
1101	Reserved
1110	Reserved
1111	Reserved

Command definitions:

- *Read*: starting at the specified address, read n sequential Q-words, where $n = (\text{length\_field} + 1)$.
- *Read (high-priority)*: This is same as Read, but the request is queued in a separate high-priority queue. The reply data can be returned out of order; however, it has to be returned within the maximum latency established for high-priority accesses.
- *Write*: starting at the specified address, write n sequential Q-words, where $n = (\text{length\_field} + 1)$. Writes have to obey the bus-ordering rules.

- *Write (high-priority)*: This is same as Write, but it indicates that there cannot be more than specified latency within which the write data has to be transferred from the master. The maximum latency is established for high-priority accesses.
- *Long read*: This is same as Read except for access size. For Long Read, the actual transfer is four times what it would have been for a normal read command with the same length field value. So, for Long Read case $n = 4 \times (\text{length\_field} + 1)$. This allows for up to 256 byte transfers.
- *Long read (high-priority)*: This is same as Read (high-priority) except for access size. The access size for Long Read is same as for Long Read. So, fundamentally, this command combines the Long Read and Read (high-priority) commands.
- *Flush*: This is similar to read. This command drives all low-priority write accesses ahead of it, standing in various queues, to the point that all the results are fully visible to all other system agents. As an indication of its completion, the command returns a random Q-Word data. The address and length fields are not used by this command.
- *Fence*: This command creates a boundary in a single master's access stream; around this boundary, writes may not pass reads.

Pin description

In addition to PCI signals or pins, there are 16 new AGP interface signals defined. The AGP-compliant target is required to support all 16 signals. These signals are ST[2::0], RBF#, SBA[7::0], AD_STB0, AD_STB1, SB_STB, and PIPE#.

AGP versions

Since AGP specification 1.0, which was defined in 1997 by Intel, there have been other revisions to AGP. Table 5.19 captures the specification and corresponding data rate.

Table 5.19 AGP Versions and Corresponding Bus Parameters

Specification	Speed	Rate (MB/s)	Frequency (MHz)	Voltage (V)
AGP 1.0	1×	266	66	3.3
AGP 1.0	2×	533	66	3.3
AGP 2.0	4×	1066	66	1.5
AGP 3.0	8×	2133	66	0.8

AGP extensions

In addition to three revisions of AGP specification 1.0, multiple official and unofficial extensions have been implemented. The main official extensions include AGP Pro and 64-bit AGP, and unofficial ones like ultra AGP, ultra AGPII, AGP express, AGI, AGX, XGP, and AGR. Complete discussion of various extensions is beyond the scope of this book.

PCI Express (PCIe)

Despite the fact that PCI was doing well, the processors and I/O devices were demanding much higher I/O bandwidth than PCI 2.2 or PCI-X could deliver; therefore, a new generation of PCI was to be engineered to serve as a standard I/O bus for future- generation platforms. PCI Express is a high-performance, general-purpose I/O interconnect defined for a wide variety of future computing and communication platforms. This is also called 3GIO, for 3rd Generation I/O interface. Key PCI attributes, such as its usage model, load-store architecture, and software interfaces, are maintained, whereas PCI's parallel bus implementation is replaced by a highly scalable, fully serial interface. PCI Express takes advantage of advances in point-to-point interconnects, switch-based technology, and packet-based protocol to deliver higher level of performance and newer features. Power management, quality of service, hot-plug/hot-swap support, data integrity, and error handling are the key new features supported by PCI Express.

PCIe design guiding principles

Being a successor of PCI, PCIe has the following guiding principles:

It should support multiple market segments and emerging applications: unified I/O architecture for desktop, mobile, workstation, server, communications platforms, and embedded devices

Ability to deliver low-cost, high-volume solutions at par or below PCI cost structure at the system level

Support multiple platform interconnect usages like chip-to-chip, board-to-board via connector or cabling

Enable new mechanical form factors: mobile, PCI-like form factor and modular, cartridge form factor

PCI-compatible software model:

– Ability to enumerate and configure PCI Express hardware using PCI system configuration software implementations with no modifications

   – Ability to boot existing operating systems with no modifications

   – Ability to support existing I/O device drivers with no modifications

- Ability to configure/enable new PCI Express functionality by adopting the PCI configuration paradigm

Performance:

- Low-overhead, low-latency communications to maximize application payload bandwidth and link efficiency
- High-bandwidth per pin to minimize pin count per device and connector interface
- Scalable performance via aggregated lanes and signaling frequency

Advanced features:

- Function-level power management along with the capability sequence device powerup sequence to allow graceful platform policy in power budgeting
- Ability to support differentiated services, that is, different qualities of service (QoS)
- Ability to support link-level data integrity for all types of transaction and packets, along with ability to support end-to-end data integrity, needed to enable high-availability solutions
- Ability to support advanced error reporting and handling to improve fault isolation and recovery solutions
- Ability to test electrical compliance via simple connection to test equipment

PCIe system architecture

The general system architecture based on PCIe is as shown in Figure 5.52. A fabric is composed of point-to-point Links that interconnect a set of components. This figure illustrates a single fabric instance referred to as a hierarchy—composed of a Root Complex (RC), multiple Endpoints (I/O devices), a Switch, and a PCI Express to PCI/PCI-X Bridge, all interconnected via PCI Express Links.

In the following section(s) we will define the terminologies used earlier.

- *Root Complex*: A Root Complex (RC) means the root of an I/O hierarchy that connects the CPU/memory subsystem to the I/O.
- *Endpoint*: Endpoint refers to a Function that can be the Requester or Completer of a PCI Express transaction. Endpoints are categorized as either legacy, PCI Express, or Root Complex Integrated Endpoints.
- *Switch*: A Switch is defined as a logical assembly of multiple virtual PCI-to-PCI Bridge devices as illustrated in Figure 5.53.

The switch replaces the multi-drop bus and is used to provide fan-out for the I/O bus. A switch may provide peer-to-peer communication between different endpoints and this traffic, if it does not involve cache-coherent memory

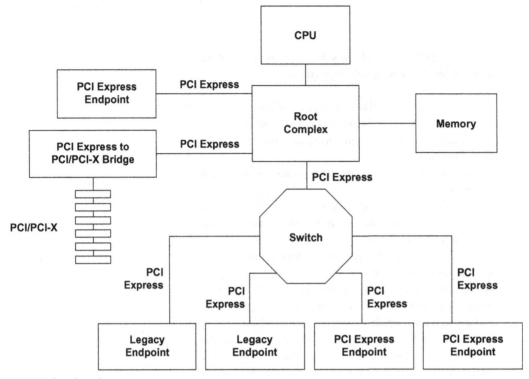

■ **FIGURE 5.52** Example topology.

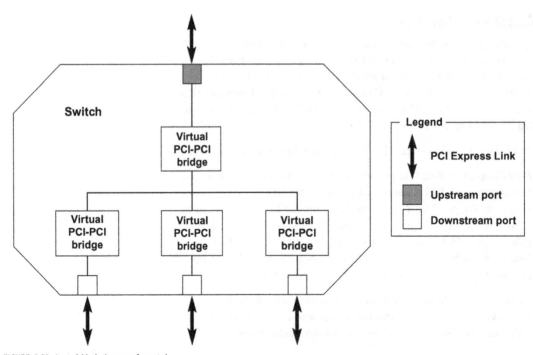

■ **FIGURE 5.53** Logical block diagram of a switch.

transfers, need not be forwarded to the host bridge. The switch is shown as a separate logical element, but it could be integrated into a host bridge component. Switches appear to configuration software as two or more logical PCI-to-PCI Bridges. A Switch forwards transactions using PCI Bridge mechanisms.

PCIe protocol architecture

The PCI Express Architecture can be specified as layers in Figure 5.54. Compatibility with the PCI addressing model is maintained to ensure that all existing applications and drivers operate unchanged. PCI Express configuration uses standard mechanisms as defined in the PCI Plug-and-Play specification. The software layers will generate read and write requests. Transaction layer transports them to the I/O devices using a packet-based, split-transaction protocol. The link layer adds sequence numbers and CRC to these packets, which makes it a highly reliable data transfer mechanism. The Physical Layer consists of a dual simplex channel implemented as a transmit pair and a receive pair. The initial speed of 2.5 Gb/s/direction provides a 200-MB/s communications channel that is close to twice the classic PCI data rate.

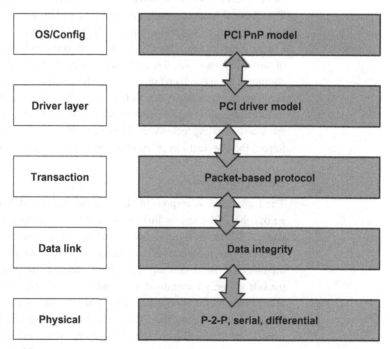

■ **FIGURE 5.54** PCIe protocol layers.

From the logical understanding perspective, the PCIe protocol architecture is built of three discrete logical layers: the Transaction Layer, the Data Link Layer, and the Physical Layer. Each of these layers is divided into two sections: one that processes outbound (to be transmitted) information and one that processes inbound (received) information. This fundamentally means that PCIe is a dual simplex protocol; that is, it can transmit and receive at the same time but it has separate channels for each of these. The Transaction Layer has data, header, and CRC to send; the data link layer adds the sequence number and LCRC; and when it goes down, the Physical Layer adds the framing information. On the receiver side the reverse is performed.

Physical Layer

The fundamental PCI Express link consists of two low-voltage differentially driven pairs of signals: a transmit pair and a receive pair. A data clock is embedded using the 8b/10b encoding scheme to achieve high data rates. The initial frequency is 2.5 Gb/s/direction and with silicon technology advances, can go up to 10 Gb/s/direction. The Physical Layer transports packets between the link layers of two PCI Express agents.

PCI Express link is composed of multiple lanes; therefore, the bandwidth of a PCI Express link can be scaled by adding signal pairs to form additional lanes. The Physical Layer supports different lane widths: x1, x2, x4, x8, x12, x16, and x32. Each byte is transmitted, with 8b/10b encoding, across the lane(s). During initialization, each PCI Express link is set up following a negotiation of lane widths and frequency of operation by the two agents involved in the communication (each end of the link). This negotiation happens without any firmware or operating system software involvement. The PCI Express architecture can support future performance enhancements via speed upgrades and advanced encoding techniques. The future enhancements, therefore, will only impact the Physical Layer, and the upper layers remain unaffected.

Data Link Layer

The Link Layer is responsible for ensuring reliable delivery of the packet across the PCI Express link. It does that by adding a sequence number and a CRC to the Transaction Layer packet. Except for packets that are retransmitted, most packets are initiated at the Transaction Layer. It uses credit-based flow control. The credit-based flow control ensures that packets are only transmitted when buffer is available to receive the packet at receiving end. This eliminates any packet retries and associated waste of bus bandwidth. Using the CRC, packet corruptions can be detected, and, transparent to other layers, the Link Layer will automatically retry a packet that has been signaled as corrupted.

Transaction Layer

The upper layer of the architecture is the Transaction Layer. The Transaction Layer receives read and write requests from the Software Layer and converts them to request packets for transmission to the Link Layer. All requests are implemented as split transactions. The Transaction Layer is also responsible for receiving response packets from the Link Layer and matching them with the original software requests. Each packet has a unique identifier that enables response packets to reach the originator. The packet format supports both 32-bit and extended 64-bit memory addressing.

The Transaction Layer supports four address spaces: the three PCI address spaces (memory, I/O, and configuration) and an additional Message Space. PCIe uses Message Space to support functionalities that were earlier provided/implemented by sideband signals, such as interrupts, power-management requests, and so on. You could think of PCI Express Message transactions as "virtual wires" since they help eliminate the wide array of sideband signals used in a platform implementation.

Software Layer

Software compatibility is of paramount importance for the newer generation I/O interconnects. There are two aspects of software compatibility: initialization, or enumeration, and runtime. PCI has a robust initialization model wherein the operating system can discover all of the add-in hardware devices present and then allocate system resources, such as memory, I/O space, and interrupts, to create an optimal system environment. The PCI configuration space and the programmability of I/O devices are key concepts that are unchanged within the PCI Express architecture. Due to this compatibility between PCIe and PCI, all operating systems will be able to boot without any modification on a PCI Express-based platform.

Performance characteristics

PCI Express's differential, point-to-point connection is able to provide high-speed interconnect using few signals. Its message space eliminates need for sideband signals, which helps minimize number of signals in implementation.

IMAGING SUBSYSTEM

As we discussed in Chapter 3, by *imaging* we refer to the system components involved in video capture. In the following sections we will talk about the interface used for interfacing a camera with the system.

Camera Serial Interface (CSI)

CSI is the most prominent interface for camera/imaging device in mobile and other low- power devices. In the following section we'll talk about the details of CSI.

Overview

The Camera Serial Interface specification (from the MIPI Alliance) defines an interface between digital camera modules and a mobile device's application processor. It has become very popular recently, and most of the camera modules for mobile devices come with CSI. The CSI specification defines standard data transmission and control interfaces between transmitter and receiver.

Data transmission interface (referred as CSI) is unidirectional, differential, serial interface with data and clock/strobe signals. The control interface (referred as CCI, Camera Control Interface) is a bidirectional control interface compatible with the I^2C standard.

Since various aspects of the CSI protocol and signaling are compatible with the DPHY (another MIPI) standard, and these aspects are described in greater detail in Chapter 4 dedicated to DSI, this chapter provides a quick overview and we assume that the reader is already familiar with DSI.

CSI layer definitions

As with the DSI, the layer definition of the Camera Serial Interface (CSI) is shown in Figure 5.55. The four fundamental layers are Phy (Physical), LML (Lane merging Layer), LLP (low-level protocol), and Application. The application layer is concerned with how the camera frame buffer interfaces with a CSI module; the buffer management, type, format, and packing of data is handled here. The LLP handles the arbitration of different request of command and data. The LML handles the byte alignments coming due to the splitting of the stream of data coming through various physical lines (DPHY).

CSI protocol

Since the CSI Physical Layer (PHY) is DPhy, most aspects of the CSI protocol and data signaling were discussed in the DSI section of Chapter 4; here, we present a brief review.

The CSI 2.0 protocol is a byte-orientated packet-based protocol. It has six components:

- Lower Power State (LPS)
- Start of Transmission (ST)

Transmitter   Receiver

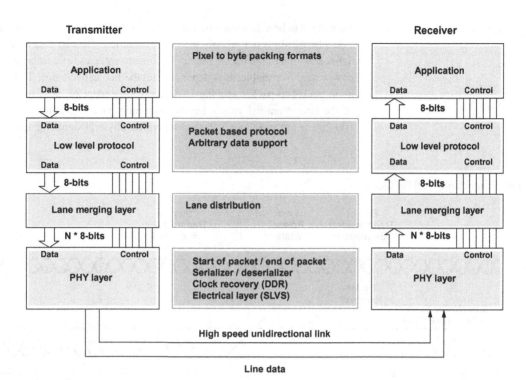

FIGURE 5.55 CSI 2.0 layer definitions.

- Packet Header (PH)
- Byte-Based Packet Data
- Packet Footer (PF)
- End of Transmission (EoT)

Lower Power State (LPS). During the low-power state (LPS), the termination resistors at the transmitter and the receiver are disabled to save power. The low-power state is signaled by a signal voltage level higher than the low-power state threshold voltage. The low-power state is a line state between packets. Data lines always have a low-power state between packets. The clock may optionally enter the low-power state between packets. The exit from LPS sequence is termed the Start of Transmission (SoT) sequence. The 8-bit leader sequence in SoT sequence is used by the receiver to synchronize to the byte boundaries within the bit-SDA streams. The SoT sequence is also used by the receiver to determine the first byte of data transmitted across the link. The entry to LPS sequence is termed the End of Transmission (EoT) sequence. The EoT sequence is used by the receiver to determine the last byte of data transmitted across the link. There is always guaranteed to be a data transition at the end of last bit of the packer footer. If a data transition is not present, then one is inserted. The last data transition defines the end of the packet footer. The time to enter and exit from the low-power state is variable. It is dependent on the line capacitance. The minimum packet-to-packet spacing is similarly dependent on the capacitance of the line. Nominal values for the minimum packet spacing are in the 100-200 ns range. Figure 5.56 shows a typical physical signaling of packets

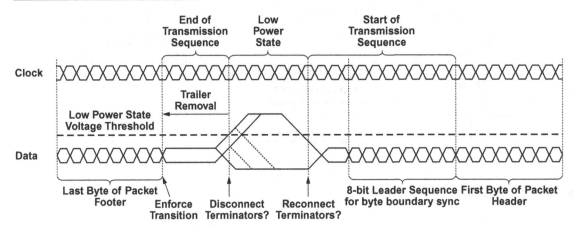

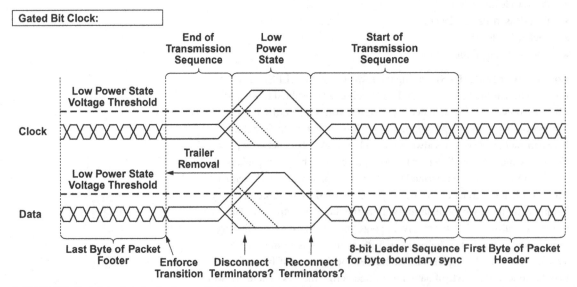

■ **FIGURE 5.56** Physical signaling of packets between Start/End of Transmission Sequences.

between Start/End of Transmission Sequences. Note that, as in DSI, CSI clock can also be continuous or discontinuous.

Packet types. As shown in Figure 5.57, there are two types of packets:

- Long Packet
- Short Packet

A long packet is composed of three elements:

- 32-bit Packet Header
- A variable number of data bytes
- 16-bit Packet Footer

A short packet is composed of only one element:

- 32-bit Packet Header

The intention of the short packet is to provide a shorter packet for encoding video synchronization information within the data stream. Both the long and short packet structures are fully specified within the low-level protocol section.

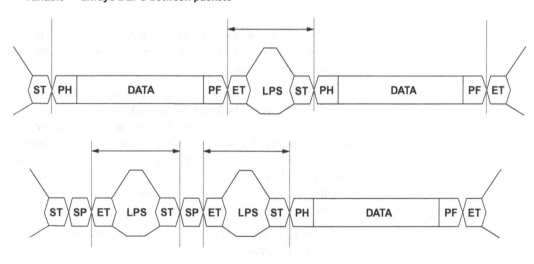

Short / long packet spacing:
Variable — always a LPS between packets

KEY:

LPS - Low Power State   PH - Packet Header

ST - Start of Transmission   PF - Packet Footer

ET - End of Transmission   SP - Short Packet

■ **FIGURE 5.57** Packet spacing.

Low-level protocol. The low-level protocol long packet structure consists of

- 32-bit Packet Header (PH)
- A variable number of 8-bit data words
- 16-bit Packet Footer (PF)

The Packet Header is composed of three elements:

- Data Identifier (DI)
- 16-bit Word Count (WC) Value
- ECC Code for the packet header

The Packet Footer is composed of one element:

- Checksum (CS)

Exit from the low-power state is followed by the Start of Transmission (SoT) sequence, which indicates the start of the packet. The data identifier defines the virtual channel for the data and the data type for the application-specific payload data. The word count value tells the receiver of length of the data line/packet. The receiver reads the word count value and then reads the next word count × 8-bit data words. During the reading of the 8-bit words, the receiver is *not* looking for any embedded sync codes, thus there are no limitations to the values of the data words, allowing arbitrary data to be transmitted.

A word count value of zero is a special case. If the word count is zero, the packet only contains the packet header; the packet footer is not present. This is the short packet format.

The Error Correction Code (ECC) byte allows single-bit errors to be corrected and 2-bit errors to be detected in the packet header. This includes both the data identifier value and the word count value. Once the receiver has read the word count × 8-bit data words, then the receiver will start looking for the next embedded sync code. In the generic case, the length of the payload data must be a multiple of 8-bit data words. However, each data format will impose additional restrictions on the length of the payload data, such as multiples of 4 bytes.

The End of Transmission (EoT) sequence followed by the low-power state indicates the end packet.

Packet spacing. Between packets there must always be a transition into and out of the low-power state. The packet spacing does not have to be a multiple of 8-bit data words as the receiver will resynchronize to the correct byte boundary during the start of transmission sequence prior to the packet header of the next packet. The time to enter and exit from the low-power state is variable. It is dependent on the line capacitance. The minimum packet-to-packet spacing is similarly dependent on the capacitance of the line. Nominal values for the minimum packet spacing are in the 100-200 ns range. Figure 5.57 shows the packet spacing of a long and short packet.

Data Packet Identifier (PI). The Packet Identifier byte contains the Virtual Channel Identifier (VC) value and the Data Type (DT) value. The Virtual Channel Identifier is the 2MS bits of the Data Identifier Byte. The Data Type is the 6LS bits of the Data Identifier Byte.

Virtual Channel Identifier (VC). The purpose of the Virtual Channel Identifier is to provide separate channels for different data flows, which are interleaved in the data stream. The Virtual channel identifier number is in the top 2 bits of the Data Identifier Byte. The Receiver will monitor the virtual channel identifier and de-multiplex the interleaved video streams to their appropriate channel. A maximum of four data streams is supported. Valid channel identifiers are 0-3. The virtual channel identifiers in the transmitters must be fully programmable to allow the host to configure which channels the different video data stream use. The principle of logical (virtual) channels is presented in Figure 5.58.

Thus various types of data defined for various types of end functions can be interleaved within a data transmission phase as shown in Figure 5.59.

Data type (DT). The data type value specifies the format and content of the payload data. A maximum of 64 data types are supported. There are eight different data-type classes (Table 5.20). Within each class there are up to eight different data-type definitions. The first two classes denote short packet data types. The remaining six classes denote long packet data types.

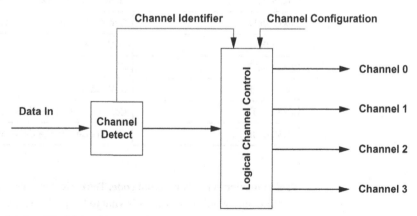

■ **FIGURE 5.58** Logical channel block diagram.

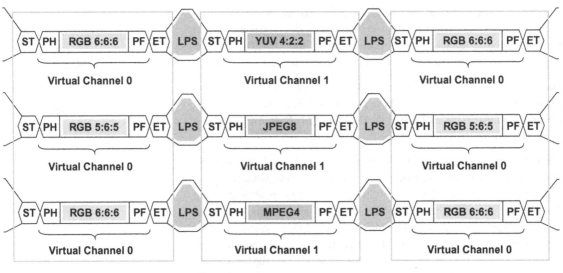

KEY:
LPS – Low Power State
ST – Start of Transmission
ET – End of Transmission

PH – Packet Header
PF – Packet Footer
SP – Short Packet

■ **FIGURE 5.59** Interleaved video data stream examples.

Table 5.20 Data Type Classes

Data Type	Description
0x00–0x07	Synchronization short packet data types
0x08–0x0F	Generic short packet data types
0x10–0x17	Generic long packet data types
0x18–0x1F	YUV data
0x20–0x27	RGB data
0x28–0x2F	RAW data
0x30–0x37	Compressed image data—stills
0x38–0x3F	Compressed image data—video

Packet header error correction code. The correct interpretation of the data identifier and word count values is vital to the packet structure. The packet header Error Correction Code byte enables 1-bit errors in the data identifier and the word count to be corrected and 2-bit errors to be detected.

Checksum generation. To detect possible errors in transmission, a checksum is calculated over each data packet. The checksum is realized as CCITT 16-bit CRC. The generator polynomial is $x16+x12+x5+x0$. The 16-bit checksum sequence is transmitted as part of the packet footer.

Frame format. The frame format of CSI will be discussed briefly here. The various components of the frame are shown in Table 5.21. For more detailed discussion please refer to Chapter 4 on DSI.

Table 5.21 Synchronization Short Packet Data Type Codes

Data Type	Description
0x00	Frame start code
0x01	Frame end code
0x02	Line start code
0x03	Line end code
0x04	Frame 1
0x05	Frame 2
0x06	Frame 3
0x07	Frame 4

Synchronization short packet data type codes. Table 5.21 lists the synchronization short packet data type codes.

Generic short packet data type codes. Generic short packet data types codes are shown in Table 5.22.

Table 5.22 Generic Short Packet Data Type Codes

Data Type	Description
0x00	Generic short packet code 1
0x01	Generic short packet code 2
0x02	Generic short packet code 3
0x03	Generic short packet code 4
0x04	Generic short packet code 5
0x05	Generic short packet code 6
0x06	Generic short packet code 7
0x07	Generic short packet code 8

It is mandatory that the short packet data type codes are transmitted using the short packet format. The intention of the generic short packet data types is to provide a mechanism for including timing information for the opening/closing of shutters, triggering of flashes, and so on within the data stream.

Packet sequence in frame transmission. It is mandatory that the short packet data type codes are transmitted using the short packet format. Each image frame begins with a short packet containing the Frame Start Code. This is termed the *frame start packet*. It is then followed by one or more long packets containing image data. Once the last long packet of the frame has been transmitted, it must be followed by a short packet containing the frame end code. This is termed the *frame end packet*. Each packet is separated by the SoT, LPS, EoT sequence.

For uncompressed image data, one long packet is equal to one line of image data. Each long packet within the same virtual channel must have equal length. For compressed image data such as JPEG, the frame is transmitted as a sequence of arbitrarily sized packets; the packet size can be tuned to minimize any buffering requirement. The period between the packet footer of one long packet and the packet header of the next long packet is called the line blanking period. The period between the frame end packet in frame N and the frame start packet in frame N + 1 is called the frame blanking period. The line blanking period may vary; the receiver should be able to cope with a near-zero line blanking period. Practical minimum line blanking periods are nominally 100-200 ns. The transmitter defines the minimum time for a frame blanking period. It is recommended that frame blanking period duration is programmable in the transmitter. The total size of data within a long packet is a multiple of 8 bits. Frame start/end packets must always be used. Figure 5.60 shows the packet position of sync in comparison to analog signaling.

Recommendations for frame start and end packet spacing:

- The frame start packet to first data packet spacing should be as close as possible to the minimum packet spacing.
- The last data packet to frame end packet spacing should be as close as possible to the minimum packet spacing.

The intention is to ensure that the frame start and end packet accurately denote the start and end of the frame of image data. A valid exception is when the positions of the frame start and end packets are being used to convey accurate vertical synchronization timing information.

The positions of the frame start/end packets can be varied within the frame blanking period in order to provide accurate vertical synchronization timing

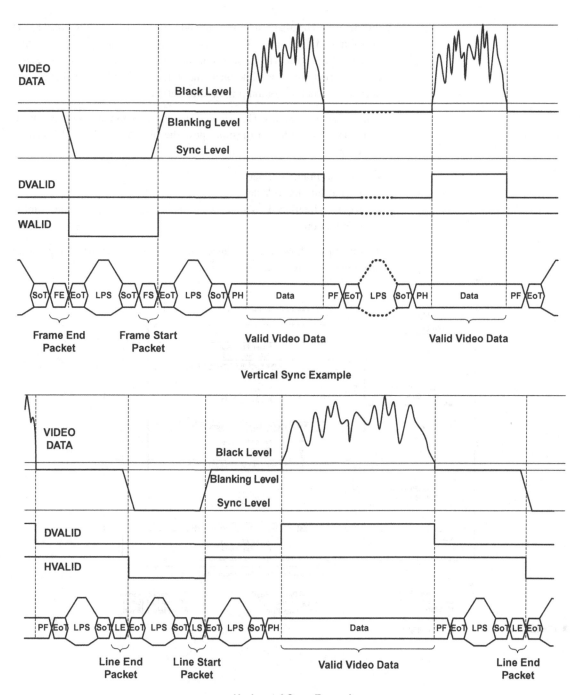

FIGURE 5.60 Example of packet position in comparison to analog signaling.

information. The use of the Line start and Line end packets is optional and is only required if the application requires accurate horizontal synchronization timing information. The positions of the line start/end packets (if present) can be varied within the line blanking period in order to provide accurate horizontal synchronization timing information.

Frame format examples. This section contains three usage examples, shown in Figures 5.61–5.63, to illustrate how the features of the CSI 2 can be used for:

- General Frame Format
- Digital Interlaced Video
- Digital Interlaced Video with accurate synchronization timing information

The examples shown are correlated with the frame signaling of interleaved or progressive signaling.

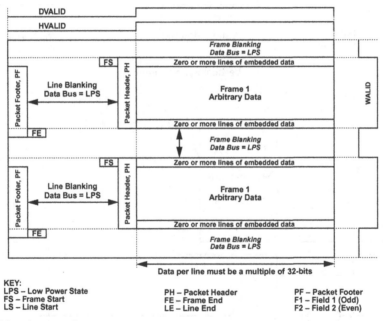

■ **FIGURE 5.61** General frame format examples.

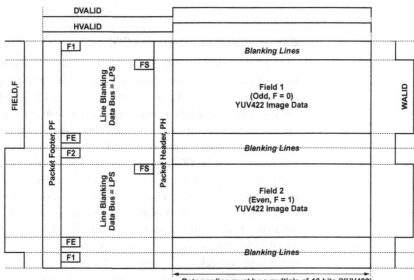

DVALID
HVALID
FIELD,F
Packet Footer, PF
F1
FE
F2
FE
F1
FS
FS
Line Blanking
Data Bus = LPS
Line Blanking
Data Bus = LPS
Packet Header, PH
Blanking Lines
Field 1
(Odd, F = 0)
YUV422 Image Data
Blanking Lines
Field 2
(Even, F = 1)
YUV422 Image Data
Blanking Lines
WALID
Data per line must be a multiple of 16-bits (YUV422)
KEY:
LPS – Low Power State
FS – Frame Start
LS – Line Start
PH – Packet Header
FE – Frame End
LE – Line End
PF – Packet Footer
F1 – Field 1 (Odd)
F2 – Field 2 (Even)

■ **FIGURE 5.62** Digital interlaced video example.

DVALID
HVALID
FIELD,F
Packet Footer, PF
Line End, LE
F1
FS
FE
F2
FS
FE
F1
Line Blanking
Data Bus = LPS
Line Blanking
Data Bus = LPS
Line Start, LE
Packet Header, PH
Blanking Lines
Field 1
(Odd, F = 0)
YUV422 Image Data
Blanking Lines
Field 2
(Even, F = 1)
YUV422 Image Data
Blanking Lines
WALID
Data per line must be a multiple of 16-bits (YUV422)
KEY:
LPS – Low Power State
FS – Frame Start
LS – Line Start
PH – Packet Header
FE – Frame End
LE – Line End
PF – Packet Footer
F1 – Field 1 (Odd)
F2 – Field 2 (Even)

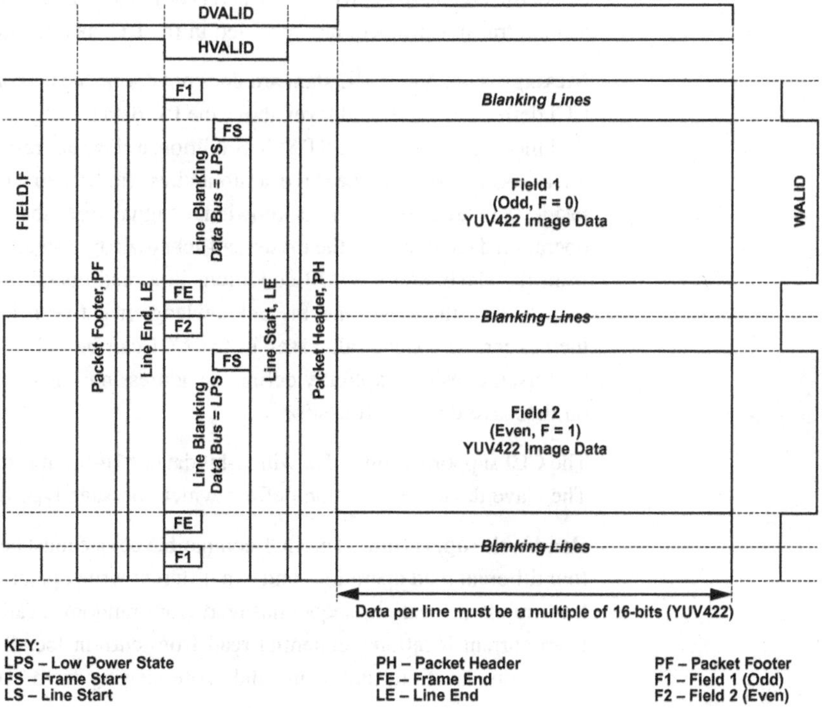

■ **FIGURE 5.63** Digital interlaced video example with accurate synchronization timing information.

Camera Control Interface (CCI)

The CCI is an I^2C fast-mode compatible interface for controlling the transmitter. To understand more about the I^2C fast-mode and its signaling, please refer to the beginning of this chapter dedicated to I^2C. The CSI receiver is always a master and CSI transmitter always a slave in the CCI bus. CCI is capable of handling several slaves in the bus, but multimaster mode is not supported. Typically no devices other than the CSI receiver and transmitter are connected to the CCI bus. This makes a pure software implementation also possible. Typically the CCI is separate from the system I^2C bus, but I^2C compatibility ensures that it is also possible to connect the transmitter to system I^2C bus. CCI is a subset of the I^2C protocol including the minimum combination of obligatory features for an I^2C slave device in the I^2C specification. Therefore transmitters complying with the CCI specification can also be connected to a system I^2C bus. Each transmitter conforming to the CCI specification may have additional features implemented to support I^2C, but that is dependent on the implementation. The CCI defines an additional layer of data protocol on top of I^2C. The data protocol is presented in the following discussions. The control interface is either CCI or system I^2C.

Data transfer protocol. The data transfer protocol is according to the I^2C standard. The START, REPEATED START, and STOP conditions as well as data transfer protocol are specified in the I^2C specification.

Message type. As the I^2C standard does not define the message formats, the CCI defines an additional layer above the I^2C data transfer protocol. A basic CCI message consists of a START condition, a slave address with read/write bit, acknowledge from the slave, a sub-address (index) for a pointing register inside the slave device, an acknowledge signal from the slave in a write operation data byte from the master, an acknowledge/negative acknowledge from the slave, and a STOP condition. In a read operation, the data byte comes from the slave and the acknowledge/negative acknowledge from the master. The slave addresses in the CCI can be either 7-bit or 10-bit. CCI itself does not explicitly define the addressing scheme; it is dependent on the slave device in question.

The CCI supports 8-bit index with 8-bit data or 16-bit index with 8-bit data. The slave device in question defines which message type to use.

Read/write operations. The CCI-compatible device must be able to support four different read operations and two different write operations: single read from random location, sequential read from random location, single read from current location, sequential read from current location, single write to random location, and sequential write starting from random location.

The index in the slave device has to be auto-incremented after each read/write operation. Typical CCI timing elements are equivalent to the corresponding I^2C bus.

Unsupported I^2C features in CCI. The differences between I^2C and CCI are as follows:

- *General call address and other "special" addresses are not supported*: general call address (0000 0000), start byte (0000 0001) transmitted as an address, CBUS and other different bus format addresses (0000 001X, 0000 010X) and the reserved addresses (0000 011X, 1111 1XXX) are not used in CCI and are therefore unsupported.
- *High-speed mode is not supported*: The CCI does not support the Hs (=high speed) mode, signified by the first address byte being (0000 1XXX).
- *Start byte is not supported*: Start condition is implied as specified in the I^2C standard. Start byte is not supported.

Chapter 6

Communication Interfaces

Following up on the general discussion of various interfaces in Chapter 3, this chapter discusses, in detail, the various interfaces used to integrate communication interfaces, their applicability to various scenarios, and their capabilities. After this chapter, the reader should understand when to use a particular interface and what advantages they have over other interfaces for a particular design.

BLUETOOTH INTERFACES

As discussed in Chapter 3, Bluetooth is a de facto standard for short-range, low-speed, low-power, wireless, point-to-point communication of control and data. A Bluetooth chip is generally not integrated as part of an system on chip (SoC), but rather mounted on the motherboard over an interface that connects the Bluetooth chip to the SoC; such an interface is referred to as a *host control interface* (HCI). Since the data transfer over Bluetooth is carried out in serial fashion, serial interfaces are the most suited for connecting a Bluetooth chip to an SoC. The most prominent of serial interfaces are Universal Asynchronous Receiver Transmitter (UART) and Universal Serial Bus (USB).

The Bluetooth chip receives the data from another Bluetooth device by wireless means and sends it to the SoC via an HCI; it also receives the data over the HCI from the SoC and transmits to another Bluetooth device. Typically, UART or USB is used as the host controller interface for a Bluetooth chip. In the following section we discuss both the UART and USB interfaces and how they are used for Bluetooth integration.

UART

UART is an acronym that stands for *Universal Asynchronous Receiver Transmitter*. The UART controller is responsible for receiving parallel data from host and transmitting that in serial fashion, bit by bit. The physical-level properties of transmission like signaling type and voltage levels are independent of the UART controller.

The UART takes bytes of data, serializes them into bits, and then transmits the individual bits in the sequence. The process is reversed at destination side. The UART controller at the destination receives the bits in sequence and then converts them into bytes for consumption by host. To accomplish the same, each UART controller implements a shift register. The shift register is used for serial to parallel (bits to byte) and parallel to serial (byte to bits) conversion.

To a host system, the UART appears as an 8-bit input and output port that it can read from and write to. Whenever the host has data to be sent, it just sends the data to the UART in byte format and whenever the UART receives data from another serial device it will buffer these data in its FIFO register that acts as a buffer until host consumes the data. The controller then will indicate the availability of data to the host through an internal register bit that the host can be polling on, or through a hardware interrupt signal.

Since transmit and receive operations can happen in parallel, separate FIFOs are required for transmit and receive operations. So, for the transmit operation, the HOST writes the data to be transmitted to the FIFO and then the UART controller sends the data bits in serial fashion bit by bit. To demarcate the start and end of the transfer, the controller adds start and stop bits to the data. In addition to the start and stop bits, the controller will also add some parity bits for error detection and control.

With respect to the ordering of bits being sent, the least significant bit is always transmitted first. And if the parity is employed, the parity bit comes last.

The idle is logic high. Each character is sent as a logic-low start bit, a configurable number of data bits (usually eight, but users can choose five to eight or nine bits depending on which UART is in use), an optional parity bit if the number of bits per character chosen is not nine bits, and one or more logic-high stop bits.

The start bit signals the receiver that a new character is coming. The next five to nine bits, depending on the code set employed, represent the character. If a parity bit is used, it would be placed after all of the data bits. The next one or two bits are always in logic-high condition and called the stop bit(s). They signal the receiver that the character is completed. Since the start bit is logic low (0) and the stop bit is logic high (1), there are always at least two guaranteed signal changes between characters.

Now, on the receiving side, the UART controller will start accumulating the data bits once the start bit is detected, and will indicate to the host (typically with an interrupt) on completion of data byte. While the interrupt is the most common method of informing the host about data availability today, it's possible, however, to do it with a bit (in register space) indicating the availability of the data to host. As part of the interrupt service routine, the host will collect the data and process it appropriately. It should be understood that if the data is not collected by the ISR in time, the data may be overwritten by the further incoming data. On regular PC systems the host may be busy with other higher priority work, and therefore it's unreasonable to expect the host to collect data as soon as interrupt arrives. And also, if the controller were to trigger an interrupt on every incoming data byte, the interrupt processing overhead would be too much for a stable system. The designers therefore chose to have multibyte FIFO to avoid data being overwritten and frequent interrupts.

Now, moving from the general description, let's talk about a real UART controller. Figure 6.1 shows an example UART controller. Here's a brief description of the various blocks in the controller:

- Register Select is used to decode the address of register that the host wants to access.
- Data Bus and Control Logic are for the host to transmit and receive data to/from UART.

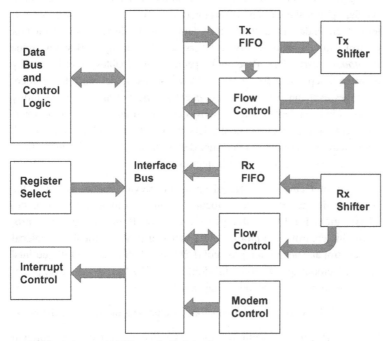

■ **FIGURE 6.1** An example UART controller block diagram.

- Internal registers: The host and the UART communicate through a set of registers. These registers function as data-holding registers (THR/RHR), interrupt status and control registers (IER/ISR), a FIFO control register (FCR), line status and control registers (LCR/LSR), modem status and control registers (MCR/MSR), programmable data rate (clock) control registers (DLL/DLH), and a user-accessible scratchpad register (SPR).
- The Shift Registers are used to break the byte information for sequential transfer or assemble the bits using FIFO to form byte data.
- The Interrupt Enable register used to enable/disable different kinds of interrupts supported by UART.
- There are other registers used to control FIFO, modem, data format, and so on.

Moving to Bluetooth, the Bluetooth core system covers the four layers and associated protocols defined by the Bluetooth specification as well as one common service layer protocol, the service discovery protocol (SDP), and the overall profile requirements specified in the generic access profile. A complete Bluetooth application requires a number of additional services and higher layer protocols defined in the Bluetooth specification.

The lowest three layers are sometimes grouped into a subsystem known as the *Bluetooth controller*. This is a common implementation involving a standard physical communications interface between the Bluetooth controller and remainder of the Bluetooth system including the L2CAP (logical link control and adaptation), service layers, and higher layers (known as the *Bluetooth host*). The Bluetooth specification enables interoperability between independent Bluetooth-enabled systems by defining the protocol messages exchanged between equivalent layers, and also interoperability between independent Bluetooth subsystems, by defining a common interface between the Bluetooth controllers and the Bluetooth hosts. In general, the Bluetooth specification does not define the details of implementations except where this is required for interoperability.

Standard interactions are defined for all interdevice operation, where the Bluetooth devices exchange protocol signaling according to the Bluetooth specification. The Bluetooth core system protocols are the radio (RF) protocol, link control (LC) protocol, link manager (LM) protocol, and logical link control and adaptation protocol (L2CAP); all of these are fully defined in the Bluetooth specification. In addition, the SDP is a service layer protocol required by all Bluetooth applications.

The details beyond this about Bluetooth are beyond the scope of the book.

When interfacing Bluetooth over the UART, the connectivity diagram will look like Figure 6.2.

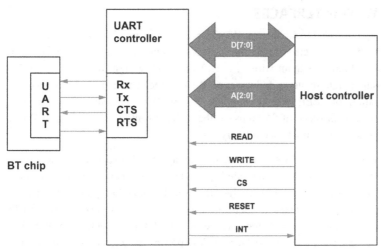

■ **FIGURE 6.2** Bluetooth over UART.

The host controller sends the data using D[7-0] channel and address using A [0-2] channel. The UART chip receives these data and using shift registers and FIFO, it converts into serial data. These serial bits will be then sent to the Bluetooth controller via Rx, Tx, CTS, RTS, and then transmitted via RF channel to other end.

USB

The USB is also used for interfacing Bluetooth chips. The USB system is made up of a host, multiple numbers of USB ports, and multiple peripheral devices connected in a tiered-star topology. To expand the number of USB ports, the USB hubs can be included in the tiers, allowing branching into a tree structure with up to five tier levels. We have already discussed USB in detail in Chapter 4. The good thing about using USB for interfacing Bluetooth is that USB is such a standard bus that the software support is already available on almost all the prevalent operating systems.

Having talked about the two primary interfaces for Bluetooth, let's do a comparative study of the two. While it is easy to use USB for interfacing a Bluetooth chip to the SoC, it's used typically while connecting an external Bluetooth dongle due to the enormous plug and play support on USB. Using UART has less overhead and less expense in terms of cost and power. UART is best suited when Bluetooth is integrated as part of platform. The UART interface cannot identify or react to a hot plug event of the Bluetooth chip and is therefore not suited for connecting external Bluetooth dongles (chips).

WI-FI INTERFACES

As discussed in Chapter 3, Wi-Fi or, more generically, wireless LAN, is a technology to enable electronic devices to connect to and communicate with the Internet without wire. It uses 2.4 GHz ultra high frequency or 5 GHz super high frequency radio waves. Wi-Fi is the general name for implementation of IEEE 802.11 standard for WLAN. The standard defines a maximum speed of 54 Mbps, so the choice of interfaces to connect Wi-Fi controllers is made accordingly.

In the majority of SoCs on the market today, the Wi-Fi chip is not integrated. In a system, the Wi-Fi module is placed on the motherboard and connected to the SoC via some I/O interface that is fast enough to support Wi-Fi usage. Earlier, PCI express (PCIe) and USB were common choices for Wi-Fi connection; however, recently secure digital input output (SDIO) has become a popular choice due to its low-power characteristics. In the following sections we will discuss these interfaces and their usage for Wi-Fi.

SDIO (secure digital input output)

SDIO is an extension of the SD (secure digital) specification designed for I/O-only devices. The SD standard was introduced in August 1999. The standard was an evolutionary improvement over the MMC (MultiMedia-Card) standard that was in use at the time. The SD standard is maintained by the SD Association (SDA). The SD format includes four card families available in three different form factors. The four families are the original Standard-Capacity (SDSC), the High-Capacity (SDHC), the eXtended-Capacity (SDXC), and the SDIO, which combines input/output functions with data storage. The three form factors are the original size, the mini size, and the micro size.

Chronologically, in 1997, Siemens AG and SanDisk developed the MMC card using Toshiba's NAND-based flash memory. Originally it used a 1-bit serial interface, but with the new architecture, it can transfer 4 or 8 bits at a time. After the release of MMC cards, most of the portable music players started using MMC cards as primary storage. However, the music industry was skeptical about the use of MMC, because MMCs would allow easy piracy of music. So, Toshiba added encryption hardware in existing MMC and named it Secured Digital or SD card. This enabled digital rights management for the music.

The SDIO card is based on and compatible with the SD memory card. This compatibility includes mechanical, electrical, power, signaling, and software. The intent of the SDIO card is to provide high-speed data I/O with

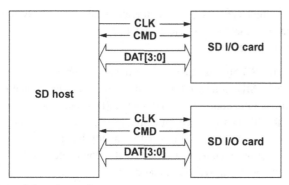

■ **FIGURE 6.3** SDIO interface signals.

low-power consumption for mobile electronic devices. Another goal is that an SDIO card inserted into a non-SDIO-aware host should not cause any physical damage or disruption of that host. In this case, the SDIO card should simply be ignored. Once inserted into an SDIO aware host, the detection of the card proceeds in regular fashion. During the normal initialization and interrogation of the card by the host, the card identifies itself as an SDIO card. The host software then obtains the card information and determines if that card's I/O function(s) are acceptable to activate.

Figure 6.3 shows the connectivity between SD host and SDIO card.

Given that the SDIO is an extension of the SD interface and is compatible with the same, we will start to investigate the SD interface first. Figure 6.4 shows connectivity of the SD card to the system.

SD bus

The SD bus is the connectivity between the host controller and SD card. Figure 6.4 shows a typical SD bus system with the host controller and the SD card. Even though the SD specification allows multiple SD cards to be connected to the same SD bus, in practical applications, most system

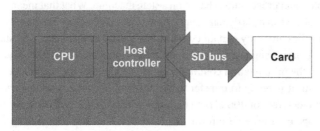

■ **FIGURE 6.4** Connectivity of SD card in a system.

designs are single card only. We will therefore talk about this system design, keeping in mind that the same can be extended to designs where multiple SD cards are connected.

Signals on the SD bus include a clock pin that is generated by the host, one bidirectional CMD (command) pin, and one or four bidirectional DT (data) pins. Figure 6.3 illustrates the signals on the SD bus interface. Communication over the SD bus is based on command and data bit streams that are initiated by a start bit and terminated by a stop bit. Before we move on, we need to define three terms that are used in the context of SD bus communication: *command, response,* and *data*:

- *Command*: A command is a token that starts an operation. A command is sent from the host either to a single card (addressed command) or to all connected cards (broadcast command). A command is transferred serially on the CMD line.
- *Response*: A response is a token that is sent from an addressed card, or from all connected cards, to the host as an answer to a previously received command. A response is transferred serially on the CMD line.
- *Data*: Data can be transferred from the card to the host or vice versa. Data is transferred via the data lines.

Each command packet consists of 48 bits of information on the CMD pin. The command packet includes the command index, argument, and cyclic redundancy check (CRC) check bits. The command is always sent by the host, and the response is sent by the card. Both the transactions traverse the CMD pin. Most response packets are also 48 bits long. As far as command initiation is concerned, the host is always the command initiator, and the card is always the command target. Not all the commands have an associated response—only read and write commands are accompanied by a data transfer. The unit of data packet is called a *block*. Different block sizes can be defined by the host; however, most data blocks are 512 bytes. Data can be transferred using either one or all four data pins. For error check, each data packet is followed by 16 bits of CRC data.

SD command processing is based on a state machine. What that means is the card is always in a particular operating state at a given time, and based on the state of the card, only certain commands are valid in that state. The state of the card may change based on the commands initiated by the host. The host, through the initialization command sequence, changes the card's operating state until it is ready to transfer data. Initially after reset, the host uses the lowest clock rate of 400 kHz to communicate with the card. Internal registers in the card provide information related to the card's capability to the host. By knowing the maximum operating frequency, data bus width, and

START	DIRECTION	COMMAND	ARGUMENTS	CRC	END

■ **FIGURE 6.5** SD command packet format.

other capability information, the host can increase the clock rate for optimal performance.

As already discussed, all SD bus transactions are initiated by the host through the CMD pin. All CMD and DT signaling are synchronized to the clock signal. Each command is a 48-bit packet and is shifted out serially by the host on the CMD pin. So, to illustrate, the generic command packet format is as shown in Figure 6.5.

The respective sizes of each field are

- Start: 1 bit
- Direction: 1 bit
- Command: 6 bits
- Argument: 32 bits
- CRC: 7 bits
- End: 1 bit

As already discussed, some commands require responses from the SD card and some do not; similarly, some commands require data transfer on the DT pins and some do not. If the command requires a response, the SD card sends out the response packet serially on the CMD pin within 64 cycles from receiving the command. The response packet format is predefined for each command. Most responses are 48 bits while some are 136 bits.

If the command has associated data transfer, it is transferred serially on the DT pins. Needless to say, the write data is transferred from the host to the card while the read data is transferred from the card to the host. Each data packet consists of a start bit, the data bits, 16 bits of CRC, and an end bit. The size of the data bit is either implicit to the command or is defined by the host with an earlier command.

The SD bus specification defines the physical signal interface on the SD bus, the definition of each command and responses, a set of standard registers within the SD card, the internal state of the card, and status of the card. The SD bus specification also defines the command sequence for initializing and enabling data transfer on the bus. It should, however, be noted that the SD bus specification does not define or restrict the type of memory or I/O devices within the SD or SDIO card.

SD host controller

As shown in Figure 6.4, the SD host controller is the hardware logic that acts as the bridge between the host CPU and the SD bus. The SD host controller specification provides a strict definition of the host controller design. Defining a standard SD host controller is intended to promote increase of SD host products that can use SD memory cards and SDIO cards. Host controller standardization enables operating system (OS) vendors to develop host drivers (SD host bus drivers and standard host controller function drivers) that work with host controllers from any vendor.

From the CPU's perspective, the host controller consists of a 256-byte register set that is mapped to the system's memory or I/O space. A transaction on the SD bus is initiated by the software reading or writing to this register set. This register set is used by host driver software to operate the card. The operation includes detection, configuration, access, and power management of the card. The SD host controller, similar to other modern controllers, has a direct memory access (DMA) to make transfers without intervention of the CPU.

The standard register map of an SD host controller is classified in 12 parts. The host controller supports byte, word, and double-word accesses to these registers. Reserved bits in all registers are fixed to zero. The host controller ignores writes to reserved bits; however, the host driver should write them as zero to ensure compatibility with possible future revisions to the specification. The host software uses these registers for operating the SD cards. Table 6.1 describes the usage for each type of registers.

Now the question is with this register set, how do we design and control a multi-slot system? The next section talks about the mechanism employed to support a multi-slot system.

Multiple slot support

On a multi-slot system, one Standard Register Set is defined for each slot. If the host controller has two slots, two register sets are required. Each slot is controlled independently.

The host driver can determine the number of slots and base pointers to each slot's Standard Register Set using the PCI Configuration register or vendor-specific methods. Offsets from 0F0h to 0FFh are reserved for the common register area that defines information for slot control and common status. The common register area is accessible from any slot's register set. This allows software to control each slot independently, since it has access to the *Slot Interrupt Status* register and the *Host Controller Version* register from each register set.

Table 6.1 Register Slots and Their Usages

No.	Register Name	Usage
1	SD command generation	Parameters to generate SD commands
2	Response	Response value from the card
3	Buffer data port	Data access port to the internal buffer
4	Host control 1 and others	Present state, controls for the SD bus, host reset, and so on
5	Interrupt controls	Interrupt statuses and enables
6	Capabilities	Vendor-specific host controller support information
7	Host control 2	Extension of host control register
8	Force event	Test register to generate events by software
9	ADMA	Advanced DMA registers
10	Preset value	Preset for clock frequency select and driver strength select
11	Shared bus	Device controls for shared bus system
12	Common area	Common information area

SDIO and SD combo cards

As discussed, the SD bus specification defines three card types: memory only card, I/O card (SDIO), and card combining memory and I/O functions (SD combo). SDIO is an extension of the SD specification designed for I/O-only devices. So, what is the difference between the SD specification and the SDIO specification? The difference is that new commands are defined in the SDIO specification while some memory-only features are removed from the base SD specification. Different I/O functions such as the SDIO Wi-Fi or Bluetooth controller can be implemented in the SDIO card format. The SDIO specification supports multifunction architecture wherein each function can operate independently and has its own memory space. The minimum number of functions is one, and the maximum is eight. Usually SDIO has two functions: function 0 and function 1.

So, what is the difference between SD combo and SDIO? It is simple: as the name suggests, the SD combo contains both SD memory and SDIO functions. It should be noted, however, that that SD combo cards are not as commonly used as SD memory or SDIO-only devices.

Standard SDIO functions

Based on the SDIO specification, there are several application specifications for standard SDIO functions. These common functions such as Wi-Fi, Bluetooth cards, and GPS receivers have a standard register interface, a common operation method, and a standard CIS (Card Information Structure) extension. The CIS extension provides additional information about the card's capability for the specific function. Implementation of the standard interfaces is optional for any card vendor, but compliance with the standard allows the use of standard drivers and applications that will increase the appeal of these cards to the consumer.

SDIO card types

The SDIO specification defines two types of SDIO cards: Full Speed and Low Speed. The Full-Speed card supports SPI, 1-bit SD, and the 4-bit SD transfer modes at the full clock range of 0-25 MHz. The Full-Speed SDIO cards have a data transfer rate of over 100 Mb/s. The second version of the SDIO card is the Low-Speed SDIO card. This card requires only the SPI and 1-bit SD transfer modes, keeping 4-bit support as optional. In addition, Low-Speed SDIO cards support a full clock range of 0-400 kHz. The intended use of Low-Speed cards is to support low-speed I/O capabilities with a minimum of hardware. The Low-Speed cards support such functions as modems and barcode scanners. If a card is a combo card, then Full-Speed and 4-bit operation is mandatory for both the memory and SDIO portions of the card.

SD card design

After discussing the fundamentals of the SD bus, now is the time to talk about a typical SD card design. By SD card, we refer to the generic SD card, which may implement either storage, I/O, or both types of functionality. The SDIO card is used for implementing Wi-Fi functionality (an SDIO card may be used for implementing other functionalities as well like GPS, modem, and so forth). Figure 6.6 shows a typical SD card design. It includes both SD memory and SDIO functions.

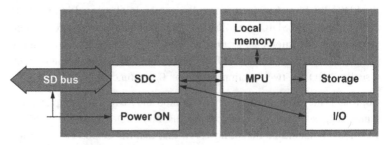

■ **FIGURE 6.6** Typical SD card design.

In Figure 6.6, SDC stands for SD controller and the Power ON block is responsible for Power ON logic. The MPU is microprocessor unit inside the card. The I/O and storage blocks represent the implementation of I/O and memory function in the combo card. The local memory is used for internal operations.

For SD memory implementation, it includes NAND flash-based storage, which is typically managed by an MPU with local memory for internal operation. The SDIO implementation includes the SD interface and the I/O function logic block. There may be an MPU inside the I/O logic block as well. The SD controller in Figure 6.6 handles all physical and data link level functions such as command decoding, response generation, CRC, status management, and a predefined SD register set. Many SD commands can be processed directly by the SD controller module without any involvement from other components in the card. However, the accesses to the memory have to be forwarded from the SD controller to other modules in the card. For example: a read or write command from a host is forwarded to the actual hardware block that is responsible for communicating with the NAND flash.

To alleviate the problem of one bad block breaking the whole card, in most NAND flash designs, access to the NAND flash is indirect and goes through a local processor that handles bad block management. The SD controller does not directly access the NAND flash chips. It either accesses a shadow memory in the local system for data transfer or interrupts the local processor, which would then process the necessary data transfer. The level of indirection helps manage the bad blocks and ensure the sanity of card. When the MPU is interrupted by the SDC, it would query the SD controller module to find out the data request from the SD bus and process that appropriately.

For an SDIO device, depending on the nature of the I/O logic being implemented, the SD controller core can either directly access the I/O logic or use the same interrupt method as described in SD memory to request service from the I/O logic. For the interrupt mechanism to be employable, an MPU is mandatory in I/O logic as well.

The architecture of the SD card shown in Figure 6.6 is more generic than specific in nature. The SD specification does not dictate the internal architecture of the SD card. The designer has the freedom to choose the architecture most suitable for the application. However, typically the trend in the industry is that the interrupt method is more commonly applied in SD memory card design, and the direct access method is more common for SDIO card design.

Electrical and timing control

Supplying the power and clock to the card is the responsibility of the host system. Typically, there is an interrupt generated to indicate hot plug and unplug events. There are multiple ways defined for implementing hot plug detection in the SD specification. Regardless of the methodologies, on receiving a card insertion interrupt, the host software can enable power to the SD card through the power control register in the host controller. In typical design, the power control information is forwarded from the controller logic to the power control unit of the system. After applying the power, the host software, via host controller, is also responsible for generating a clock signal to the SD card. After power is enabled, host software will enable the clock signal to the SD card at the default frequency of 400 kHz. The initialization sequence is run at this frequency until the software finds the frequency and bus width capability of the card through the card's control registers. The host software then may change the frequency to a more appropriate value based on the discovery of the card capability and usage requirement. In order to save power, the host CPU can stop the clock signal and/or remove power to the card when there is no operation.

Software operation

Having talked about the hardware components of the SD infrastructure, before closing, let's quickly touch on the software components in the setup.

Figure 6.7 shows the abstracted view of the SD card setup. The SD card end of the setup has an MPU as discussed earlier. The MPU will run software responsible for carrying out the commands from the host controller sent via the SD bus. This software is also responsible for doing internal management like bad blocks. The software running on the MPU will typically be known as *card firmware*. In order to enable the SD bus in the system, the OS must be aware of the SD, so there is a need for a bus driver (in Windows driver terminology). The bus driver is responsible for detecting the card existence and running through the initialization sequence. As of this writing,

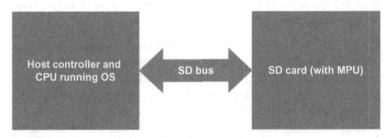

■ **FIGURE 6.7** SD system view from a software design perspective.

major operating systems have native support for bus driver operations. Additionally, the operating systems generally have built-in support for the SD card function as a memory device. In the case where the SD card is not configured as a standard SD memory device, specific application software and device drivers will be needed to communicate with the card to perform the desired functions. This device driver needed for specific desired functions is called the *function driver*.

PCI express (PCIe)

Given the speed requirement of Wi-Fi, PCIe is also a suitable interface to be used for connecting Wi-Fi chips onto the system. PCIe is software compatible with PCI and has been around for years. The software support for PCIe-based components is mature across all major operating systems. It is reasonable to expect built-in software support for all classes of PCIe devices in operating systems. PCIe-based Wi-Fi solutions are no exception. The only concern in using PCIe for Wi-Fi connectivity is that typically the PCIe is considered a power-hungry interface and therefore avoided in low-power SoC-based systems, as far as possible. However, the newer power management schemes have narrowed the difference in power consumption by a significant amount, and therefore PCIe is becoming a viable interface in low-power SoC designs as well. The PCIe interface was discussed in detail in Chapter 5.

USB

Universal Serial Bus, or USB, is also used for connecting the Wi-Fi chip. Similar to other components, the good thing about using USB for interfacing Wi-Fi is that the USB is such a standard bus that the software support is already available on almost all the prevalent operating systems. Since the USB bandwidth is shared across all the devices connected on the bus, based on the system design and configuration, the overall USB speed may be a limiting factor. USB 3.0 has however gotten rid of that concern.

In general, due to USB's plug and play capability and software support, USB is the preferred choice while connecting a Wi-Fi chip externally to the system (for example in the case of connecting a Wi-Fi dongle).

Summary

As we conclude the section, it is worth mentioning that SDIO, the same standard that is used for removable flash storage in cameras, has been the preferred choice for Wi-Fi connectivity in low-power SoC systems. USB continues to be the primary choice for external Wi-Fi modules. In the absence of SDIO, PCIe has been the interface of choice for connecting a Wi-Fi chip on the system. After SDIO became popular and the focus on lower power grew, the market started to shift toward SDIO.

The SDIO interface itself is capable of a maximum bandwidth of 100 Mbps. However, with the SDIO software stack overhead, the practical throughput of the SDIO interface can be in the 30-40 Mbps range. With the newer Wi-Fi standard proposing to support higher speeds, the SDIO throughput seems to be a bottleneck. And therefore PCIe is becoming the preferred choice once again.

One solution to this problem is a transition to a more modern interface: HSIC or SSIC. HSIC (High-Speed Inter-Chip) is a serial interface based on USB 2.0 that is destined to replace SDIO. HSIC should guarantee speeds of up to 480 Mbps (theoretically). The HSIC software stack is also apparently more robust and is not expected to pose a problem in the way that SDIO has thus far. Given that smartphones and tablets are expected to use single stream 802.11ac at 433 Mbps, HSIC looks to be the appropriate SoC interface.

In parallel to HSIC, SSIC is an interface based on USB 3.0 that is even faster than HSIC and can be used in the future, if the revised standard asks for more bandwidth.

2G/3G/4G INTERFACES

As discussed in Chapter 3, mobile telecommunication is accomplished over cellular networks. The mobile telecommunication evolution is classified as 2G, 3G, 4G, and so on. There were intermediate generations as well, like 2.5G and 3.5G. From the SoC integration perspective the fundamentals have not changed across these generations. The changes in the SoC integration strategy are primarily due to the throughput requirement changes across these generations: For example, we know 3G provides faster speed than 2G, and 4G is faster than 3G, and so on. Figure 6.8 shows the logical view of the telecommunication ecosystem.

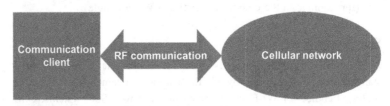

■ **FIGURE 6.8** Logical view of the telecommunication world.

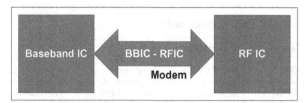

■ **FIGURE 6.9** Logical view of a modem.

Except for the very first generation of cellular network, which was analog, the cellular telecommunication has been digital. At the communication client end (such as the mobile phone), there are two key components: modem and SIM (subscriber identity module).

The modem is responsible for modulating and demodulating the signal before transmission and after reception. The modem can further be viewed as having two key components: baseband IC and RFIC. Baseband IC is responsible for processing of the data/signal within the modem, while RFIC is responsible for RF communication (see Figure 6.9).

The SIM card is a kind of smart card that stores the international mobile subscriber identity and the related key used to identify and authenticate subscribers on mobile telephone devices. Additionally it keeps data for ciphering information, and temporary information related to the local network the user is connected to, and a list of the services the user has access to on the network. It must be noted that the role of the SIM card is only to provide the metadata about the user and nothing more. The SIM card does not have a role to play in actual communication of data.

The SIM card interfaces are standardized so the SIM can be seamlessly used across devices. The SIM card interface usually has six pins. From the standard perspective, there actually are eight pins, two of which are not used. Table 6.2 provides a brief description of the connector pins.

It is important to note that the SIM card interface and design do not really depend on the generation of communication technology used. So, they can be the same across 2G, 3G, and 4G, and so on. The actual data stored in the SIM can be different, of course, and that is what really matters.

The next part in the communication client is modem, which has both baseband and RF chip. The modem connectivity interface has changed significantly due to the change in throughput requirements from generation to generation.

Table 6.2 SIM Card Connector Pins

Pin	Name	Description
1	VSIM	+5 VDC power supply input
2	SIMRST	Reset signal, used to reset the card's communications
3	SIMCLK	Provides the card with a clock signal, from which data communications timing is derived
4	RESERVED	AUX1, optionally used for USB interfaces and other uses
5	GND	Ground (reference voltage)
6	NC	Programming voltage input (optional). This contact may be used to supply the voltage required to program or to erase the internal nonvolatile memory
7	SIMDATA	Input or Output for serial data (half-duplex) to the integrated circuit inside the card
8	RESERVED	AUX2, optionally used for USB interfaces and other uses

So, initially, the UART/RS-232 was good enough to interface modems. However, later we moved on to use USB, PCIe, HSI, HSIC, SSIC, DIgRF, and so on. We will quickly go over them in the following section.

Similarly, from the cellular/RF communication standpoint, there have been many changes to enable the faster speed; for example, a variety of multiplexing technologies have been developed, such as FDMA, TDMA, CDMA, and their variations.

RS-232

RS-232 is a standard for serial communication transmission of data. This can be considered as the specific implementation of the generic UART specification. It formally defines the signals connecting between a data terminal equipment such as a computer terminal, and a DCE (data circuit-terminating equipment) such as a modem. The DCE was originally called data communication equipment. The RS-232 standard is commonly used in computer serial ports. The standard defines the electrical characteristics and timing of signals, the meaning of signals. Additionally it also defines the physical size and pin-out of connectors.

Earlier, the RS-232 serial port was a standard feature of a personal computer. The interface at that time was used to connect modems, printers, mouse, data storage, uninterruptible power supplies, and other peripheral devices. However, RS-232 has severe limitations in today's context. Some of these limitations are low transmission speed, low noise immunity, large voltage swing causing high power consumption, and large standard connectors. Additionally, the RS-232 did not define any method for sending power to a device. In modern personal computers, USB has displaced RS-232 in most of its peripheral interface roles.

USB

USB, or Universal Serial Bus, is serial protocol supporting the hot plugging of devices of multiple classes. USB immediately replaced RS-232 as the connection medium for serial devices, except for the places where the legacy was too much to overcome—modem was not an exception in this case. USB apparently addressed the key limitations of RS-232 like low transmission speed, low noise immunity, large voltage swing causing high power consumption, large standard connectors, and so on. Additionally, USB supported plug and play. USB also defined various classes of devices, and the operating systems implemented native support for various classes of the devices. So, if the end devices followed and implemented the standard properly, the device would work without any additional software support.

The USB interface and variations of USB were discussed in detail in Chapter 5.

PCIe

Due to increasing speed requirement for the modem interface, PCIe is also a suitable interface for modem integration. PCIe has a bigger connector and therefore is not suited when used externally (like a dongle) for mobile devices. However, it is a suitable interface when the modem device is integrated on the PCB. Even though PCIe can support high bandwidth, PCIe is generally considered a power-hungry interface. It is therefore used when and if real high throughput is required that cannot be supported by other lower power interfaces.

The PCI interface and its variations were discussed in detail in Chapter 5.

HSIC and SSIC

Even though the generic USB interface was good for modems, the generic physical layer interface that supports connectivity via long cables and so on can be optimized for better power and performance for inter-chip communication. The HSIC and SSIC interfaces are variations of USB 2.0 and USB 3.0, respectively, for inter-chip communications.

HSIC

USB is the ubiquitous peripherals interconnect of choice for a large number of computing and consumer applications. Many systems provide a comprehensive set of drivers to support all commonly available USB peripherals. As a result of its popularity, it was very attractive to use USB for chip-to-chip interconnection within a product, as well. When it is known that the interface in question will be used as connection between two ICs only, power, area, and functionality can be potentially optimized. HSIC is the chip-to-chip variant of USB 2.0 that eliminates the need for conventional USB PHY. The HSIC PHY uses about 50% less power and 75% less area than traditional USB 2.0 PHY. HSIC uses two signals (STROBE and DATA) at 1.2 V and has a throughput of 480 Mbps using 240 MHz DDR signaling. Maximum PCB trace length for HSIC is 10 cm. It does not have low enough latency to support RAM sharing between two chips. USB HSIC is 100% host driver-compatible with traditional USB cable-connected topologies. Full-speed and low-speed USB transfers (as in conventional USB) are not supported by the HSIC interface.

The key characteristics of HSIC are:

- High-speed 480-Mbps data transfer rate
- Source synchronous serial interface
- No power consumed unless there is a transfer in progress
- Maximum trace length on the PCB is 10 cm
- Hot plug/unplug not supported
- Signals driven at 1.2-V standard LVCMOS level
- Designed with low-power applications in mind
- The HSIC interface always operates at high speed

HSIC and standard USB comparison

HSIC is an interface that has been designed to replace a standard USB PHY and USB cable with an interface that is optimized for circuit board layout. On conventional USB 2.0, signal was converted from digital to analog, transmitted over a wire, and converted back from analog to digital in the USB PHY on the other SoC. With the new Hi-Speed Inter-Chip (HSIC) standard created by the HSIC USB-IF Working Group, signals can stay digital, transmitted to the PCB over standard traces, and be received at the destination. Eliminating the analog circuitry of the USB 2.0 PHY is the essence of the USB 2.0 HSIC PHY. For comparison see Figures 6.10 and 6.11, taken from the specification, one with standard USB and other with USB HSIC interface.

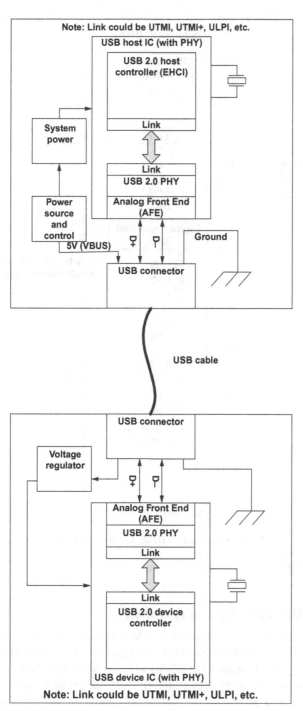

■ **FIGURE 6.10** Standard USB host and peripheral example.

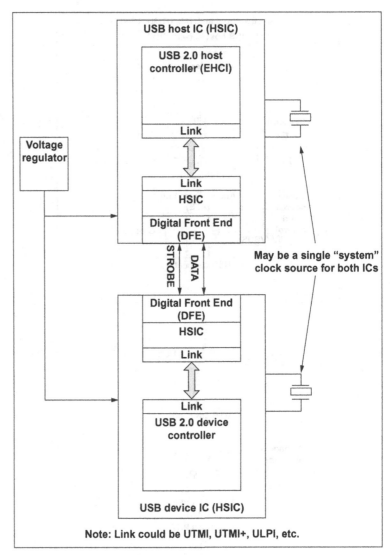

■ **FIGURE 6.11** HSIC USB host and peripheral example.

HSIC USB signaling and operation

An HSIC interface is built on two signals: a bidirectional data strobe signal (STROBE) and a bidirectional DDR data signal (DATA). The DATA signal is synchronous with the STROBE signal of the transmitter. HSIC as is utilizes the tiered-star topology of the USB 2.0 specification; therefore, the interface is a point-to-point connection between a downstream-facing host/hub port and an upstream-facing peripheral port. The data transferred includes sync, bit stuffing, EOP (end of packet), and NRZI encoding to ensure that the packet length, turnaround time, inter-packet gap, and the like are as near to identical as possible between HSIC and standard USB 2.0 signaling. This enables a single host to drive standard USB and HSIC USB interfaces without any modifications to the internal logic structure of the host.

SSIC

Because the throughput of USB 2.0 and HSIC were proving inadequate, the USBIF introduced USB 3.0 with super speed support. While bringing increased speed and throughput with USB 3.0, the USB-IF replaced HSIC with the SuperSpeed Inter-Chip (SSIC) standard for on-PCB communication. SSIC is the chip-to-chip variant of USB 3.0. The relationship between USB 3.0 and SSIC is similar to the relationship between HSIC and USB 2.0, discussed in the previous section. SSIC consumes significantly less power than the conventional super speed USB 3.0. In order to reduce power, in SSIC, the USB 3.0 PHY is replaced by MIPI M-PHY. M-PHY can come in three speeds, called *gears*. Gear1 operates at 1.25 or 1.45 Gbps, Gear2 at 2.5-2.9 Gbps, and Gear3 up to 5.8 Gbps. In addition, M-PHYs can have one, two, or four lanes. Each lane has n pins; so two lanes have $2n$ pins and four lanes have $4n$ pins. These lane configurations offer flexibility to run either in multiple parallel lanes at slower clock speeds to save power, or to run at faster speeds but consume fewer pins. Figure 6.12 shows the blocks in SSIC communication.

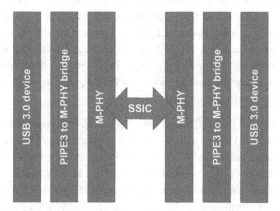

■ **FIGURE 6.12** Standardized SSIC interface between USB 3.0 and MIPI M-PHY.

Since two different standards (USB 3.0 and M-PHY) were coming together, it was important to standardize the interface between the USB 3.0 controllers and MIPI M-PHY. The MIPI alliance and USB-IF worked together to standardize the interface between the USB 3.0 controllers and MIPI M-PHY. The standard chose to preserve the PIPE interface used between the USB 3.0 controller and USB 3.0 PHY. However, there was still an interface needed to reach to M-PHY. In M-PHY standard v2.0, the SSIC interface to M-PHY is called RMMI (Reference M-PHY Module Interface). The logic bridge between the USB 3.0 controller and M-PHY is called the PHY adapter. The adapter needs to support USB 3.0 power-saving modes for all the lane/gear configuration of M-PHY. Similar to the fact that the HSIC PHY is smaller than a USB 2.0 PHY, the MIPI M-PHY is about 50% smaller than a USB 3.0 PHY. Eliminating the USB 3.0 PHY saves half of the area. In addition, a MIPI M-PHY consumes significantly less power, especially in Gear1, one-lane operation. In this configuration, a MIPI M-PHY consumes only 20% of the power of a USB 3.0 PHY. The power savings in SSIC comes from two factors: the smaller PHY and slower data rate in the Gear1, one-lane M-PHY configuration, which is only 1.25 or 1.45 Gbps. Since USB 3.0 PHY always operates at 5 Gbps, the M-PHY allows for the lower data rate and power savings.

The high-speed MIPI M-PHY, working in conjunction with USB 3.0 SSIC, is designed keeping mobile applications in mind and is becoming a popular physical layer solution. With up to 5824 Mbps bandwidth, the High Speed Gear3 serialization speed meets devices' high bandwidth requirements. The M-PHY is designed to accommodate the intermittent nature of inter-chip communications and employs burst operation to toggle between data transmission and power-saving states, effectively reducing power consumption.

As a general note, there is also a provision to multiplex M-PHY for other functions, in addition to SSIC operations. For example, a single M-PHY can be used with USB and an LLI (Low Latency Interface) controller to allow the baseband of a wireless device to interface RAM with the applications processor. By multiplexing a single PHY with two digital controllers, designers save the area of an extra M-PHY, as long as only one digital controller is working at any one time.

A fully integrated USB 3.0 SSIC controller and M-PHY enables low-power, efficient connectivity on a PCB between a smartphone/tablet application processor and a modem or Wi-Fi chip. Using an existing USB 3.0 software stack with the low-power capabilities of the MIPI M-PHY enables designers to meet the increasing performance and battery life requirements of mobile or low-power electronics.

HSI

HSI (High Speed Synchronous Serial Interface) is a high-speed communication interface. HSI was created in 2003 and is now managed by the MIPI Alliance. It is the ancestor of mobile phone inter-chip interconnects and is still present on many modern-day SoCs. HSI operates at 1.2 or 1.8 V and has throughput of 200 Mbps. HSI is a very popular interface for connecting modems to SoCs. HSI supports full-duplex communication over multiple channels and is capable of reaching speeds of 200 Mbps. It offers a high-speed multichannel interface to connect 3G and 4G (up to LTE or Long Term Evolution Category 3) modems.

HSI provides an on-chip interconnect with a low-latency communication channel over a die-to-die link by dividing the physical link to logical channels at the hardware level. HSI enables versatile, easy, and simple serial link realizations. As such, HSI can reduce time-to-market and design cost of mobile handsets by simplifying serial die-to-die physical layer implementations.

As can be seen from the connectivity diagram (Figure 6.13), the HSI protocol works on four signals: WAKE, READY, FLAG, and DATA. The WAKE signal is used by transmitter to wake up the receiver; the receiver acknowledges by the READY signal. The actual data is transmitted over

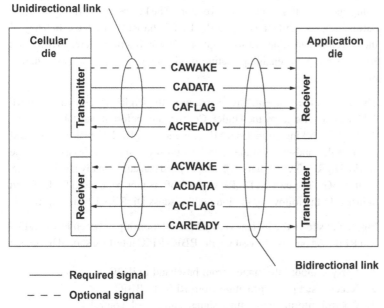

■ **FIGURE 6.13** HSI connectivity.

the DATA channel. The clock is synchronous between the application die and cellular die (or in other words transmitter and receiver) and is not separately provided. One may be wondering about FLAG. FLAG works in tandem with DATA to indicate the bit boundaries (makes up for clock signal, which is missing in this case). So, for example, if the consecutive data bits remain the same, the FLAG will toggle. Thus, between the two, only and exactly one will toggle. This kind of toggling scheme reduces the number of toggles and also saves power as a consequence.

DigRF

The initial version of DigRF for usage in GSM/EDGE handsets was specified in 2004 by the DigRF Consortium group members. Later the DigRF standard was taken over by the DigRF working group formed by MIPI alliance. The intent of the DigRF working group was to provide a high-speed scalable interface between baseband IC (BBIC) and RFIC. The working group has taken a two-pronged strategy: first to define specification for use in 4G and LTE and then for future generations.

Since the Digital Serial Interface replaces the analog interface in previous-generation mobile handset architectures, the primary purpose of DigRF is to standardize the interfaces between RFIC and BBIC to ultimately produce "plug and play" ICs from various vendors. The key benefits driving the standardization of interface between the BBIC and RFIC are the efficiency of the development time, various options in combining different BBIC and RFIC as required, enabling feature-on-demand service in a short amount of time.

There are, at the moment, three versions of the DigRF specification. DigRF 1.12 was developed by the DigRF Consortium, before it was taken over by the DigRF working group from MIPI alliance. DigRF 1.12 was used in GSM and EDGE handsets, and the specification was publicly released to all the world. DigRF v3.09 with its 312-Mbps speed can additionally handle UMTS (that is, 3G). The current DigRF v4 draft offers Gbs bandwidth for LTE and WiMax. In the following section we'll discuss DigRF 1.12 and DigRF 3.x.

Figure 6.9 shows the logical view of the relationship between baseband IC and RFIC. As one would assume, the BBIC-RFIC interface should be able to

1. Carry symbol information from baseband to RFIC
2. Receive sample information from RFIC to BBIC
3. Control information, timing signals, and clock control
4. Clock signal

There is a strong motivation (as usual) to minimize the number of IC pins needed to support the interface, especially at the RFIC end. The only way to achieve that would be to multiplex as many signals as possible without compromising the functionality. After all the possible multiplexing, the result is an interface (for DigRF 1.12) that requires eight pins. So, the basic physical layer of this standard is implemented with an eight-wire interface. Payload data is contained on a bidirectional interface (due to multiplexing), RxTxData with an accompanying RxTxEn signal, and a clock signal SysClk. The SysClk can be gated by SysClkEn. Control data (CtrlData) is a separate line with a dedicated control clock (CtrlClk) and CtrlClkEn line. A strobe line is typically used for over-the-air timing.

Figure 6.14 shows the connections.

Moving on to DigRF v3.x, this standard supports a variety of 2-2.5G and 3GPP air standards. The basic physical layer of the DigRF v3 standard is implemented with a six-wire interface. In DigRF v3, the focus has changed from multiplexing Tx and Rx, and instead control and data multiplexing have been employed. Independent transmit (TxData) and receive (RxData) differential signal pairs allow for bidirectional communication. The same signal is used for both control and IQ/payload data between the RFIC and the BBIC. SysClk and SysClkEn provide the timing references for digital serial data transmission and recovery. However, there is no strobe signal for over-the-air timing; this is accomplished via frame communication.

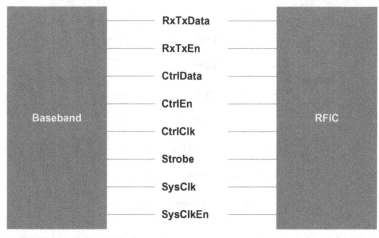

■ **FIGURE 6.14** DigRF v1.12 interface.

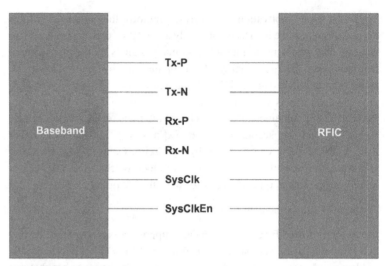

■ **FIGURE 6.15** DigRF v3 interface.

Figure 6.15 shows the connection.

DigRF v4 is intended to be used in mobile terminals that support next-generation mobile broadband technologies such as Long Term Evolution (LTE) and Mobile WiMax. It also supports existing 3GPP standards such as 2.5G and 3.5G. Due to its large data rate and scalability it is suitable to cover other non-3GPP air interfaces. All the standards discussed are supported over a common interface. Figure 6.16 shows the connectivity diagram of the DigRF v4 interface.

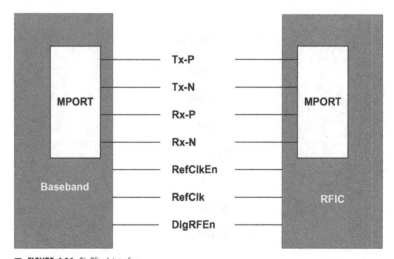

■ **FIGURE 6.16** DigRF v4 interface.

DigRF v4 focuses on the protocol and the programming model as well as parts of the physical interface between the BBICs and RFICs. It does not specify anything within either IC, except for the minimum necessary to ensure compatibility and interfaceability at these layers.

The guiding principles for DigRF v4 design are increased bandwidth, power consumption, lower pin count, minimizing the RF spurious problem, and flexibility and scalability to allow multiple transmitter and receivers. For a real example of the BBIC and RFIC using the DigRF interface, one can refer to Synopsys design incorporating MIPI DigRFv4 interface.

Summary

As we covered multiple potential interfaces to be used for connecting telecommunication modems to the systems, the key driver for change in interface has been the throughput requirement. When there are multiple options that can support the bandwidth requirement, we try to choose the one having lower power consumption. And for the competing standards for power consumption, we chose the one that is easy to integrate in terms of hardware and software availability and maturity.

Given that, overall, due to their flexible nature, strong and mature software support and low-power, HSIC and SSIC are the interfaces for modem connectivity in the present and future.

GPS INTERFACES

As discussed in Chapter 3, GPS is one of the most amazing engineering inventions in decades. Many years of engineering went into getting an accurate position anytime, anywhere. A number of GPS satellites, all containing extremely accurate atomic clocks, have been launched since the late 1970s, and launches continue even today. The satellites continuously send data down to earth over dedicated radio frequencies that are received by tiny processors and antennas on GPS receivers and compute the position and time on the fly. Before getting into the details of how the GPS chips are integrated into the system, let's quickly talk about how GPS works, describe GPS fundamentals, and then move to the GPS chip integration.

How GPS works

GPS receivers use a constellation of satellites and ground stations to compute position and time almost anywhere on earth. At any given time, there are at least 24 active satellites orbiting the earth. In fact there are 27 satellites, of which 24 are active at any moment and the remaining 3 act as backup, in case one of the 24 active satellites fails.

Each of these solar-powered satellites circles the globe at a distance of about 12,000 miles (19,300 km) above the earth, making two complete rotations every day. The orbits are arranged such that at anytime, anywhere on earth, there are at least four satellites "visible" in the sky.

The primary purpose of these visible satellites is to transmit information back to earth over radio frequency (ranging from 1.1 to 1.5 GHz). A GPS receiver's job is to locate four or more of these satellites, figure out the distance to each, and use this information to deduce its own location. This operation is based on a simple mathematical principle called trilateration.

Fundamentally, in order to make the calculations, the GPS receiver has to know two things:

- The location of at least three satellites (preferably four for more precision)
- The distance between receiver and each of those satellites

The GPS receiver figures both of these things out by analyzing high-frequency, low-power radio signals from the GPS satellites (remember the GPS satellites are always transmitting data). There could be multiple receivers, so they can pick up signals from several satellites simultaneously. The speed of radio waves is known; that is, the speed of light (about 186,000 miles per second or 300,000 km per second, in a vacuum). The receiver can figure out how far the signal has traveled by timing how long it took the signal to arrive to it.

So, how does the receiver figure out the time taken for the signal to arrive to it? The procedure is as follows. At a fixed particular time the satellite begins transmitting a long digital pattern called a pseudo-random code. The receiver begins running the same digital pattern also exactly at the same fixed time. When the satellite's signal reaches the receiver, its transmission of the pattern will lag a bit behind the receiver's playing of the pattern.

The length of the delay is equal to the signal's travel time. The receiver multiplies this time by the speed of light to determine how far the signal traveled. In order to make this measurement, the receiver and satellite both need clocks that can be synchronized down to the nanosecond. However, there are tricks applied such that receivers have no need of expensive high-precision clocks—the GPS satellites, of course, do. The idea is that the receiver looks at incoming signals from four or more satellites and corrects its own inaccuracy. Fundamental to the correction algorithm is that the correct time value will cause all of the signals that the receiver is receiving to align at a single point in space. After correction, the receiver readjusts its clock.

When the receiver measures the distance to four located satellites, it can draw four spheres that all intersect at one point (the process of trilateration). Three spheres will intersect even if the numbers are incorrect, but four spheres will not intersect at one point if the measurements are incorrect. Since the receiver makes all its distance measurements using its own built-in clock, the distances should all be proportionally incorrect. In order for the distance information to be of any use, the receiver also has to know the satellites' actual positions. Since the satellites travel in very high and predictable orbits, the GPS receiver simply stores an almanac that tells it where every satellite should be at any given time. The almanac values are static for a long period. They are transmitted by satellite and updated by the receiver on a regular basis, however. The ephemeris is the precise satellite orbit at the moment. The ephemeris is also part of the GPS message.

Now, in practice, there may be inaccuracies in determining the local position, due to the receiver's position, bad signal strength, or bad weather. A number of ways have been devised to overcome these problems.

Differential GPS

The basic idea of differential GPS (DGPS) is to gage GPS inaccuracy at a stationary receiver station with a known location. Since the DGPS hardware at the station already knows its own position, it can easily calculate its receiver's inaccuracy. The station then broadcasts a radio signal to all DGPS-equipped receivers in the area, providing signal correction information for that area. So, for the DGPS to work, the GPS receivers must receive signals from GPS satellites as well as the DGPS station transmitter.

Assisted GPS

Assisted GPS (AGPS) is a system that uses the assistance of network data to significantly improve the startup performance, or *time to first fix* (TTFF), of a GPS-based system, especially in case of poor GPS signal scenarios.

Table 6.3 shows the format of the GPS messages. The message has 5 subframes of 10 words each. The message carries the satellite clock value, ephemeris, and almanac, which are used by the GPS receivers.

In summary, the overall operation is: GPS chip gets the data from multiple satellites, does calculations, and supplies the information to the host. Due to this nature of operation, it's more logical to use a serial interface for connecting the GPS receiver chip to the host processor. The serial interfaces

Table 6.3 Format of GPS Messages

Subframes	Words	Description
1	1-2	Telemetry and handover words (TLM and HOW)
	3-10	Satellite clock, GPS time relationship
2/3	1-2	Telemetry and handover words (TLM and HOW)
	3-10	Ephemeris (precise satellite orbit)
4/5	1-2	Telemetry and handover words (TLM and HOW)
	3-10	Almanac component (satellite network synopsis, error correction)

typically used for GPS chip integration are USB and UART. Therefore GPS connectivity diagram on a system would look like Figure 6.17.

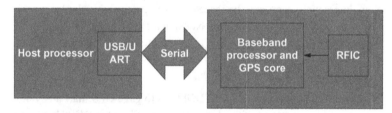

■ **FIGURE 6.17** GPS chip and host processor connectivity.

Now, having discussed the GPS module functionality and integration in general, we will briefly talk about the UART and USB interfaces.

UART

UART, as already discussed in the earlier section of the chapter, is one of the prominent low-power serial interfaces used in the industry. The UART controller is responsible for receiving parallel data from host and transmitting that in serial fashion, bit by bit. The physical-level properties of transmission like signaling type and voltage levels are independent of the UART controller. For interfacing a GPS module with a UART interface, the Rx of UART controller side is connected to the Tx of GPS and vice versa.

USB

USB is also used for interfacing GPS chips. Similar to other components, the good thing about using USB for interfacing GPS is that the USB is such a standard bus that the software support is already available on almost all the prevalent operating systems.

Summary

As we know the UART controller cannot detect a device hot plug event, neither can it identify the capabilities of a device that is being connected to it, and therefore UART is most suited when the GPS module is integrated on the PCB itself. Contrasting that, the USB, due to its plug and play capabilities and dynamic detection of device class and functionality, is ideally suited for connecting external USB GPS used as a dongle.

NFC INTERFACES

NFC, an acronym for near field communication, is a form of short-range wireless communication. The antenna in the case of near field communication is much smaller than the wavelength of the carrier signal. This is to prevent a standing wave from developing within the antenna, and so in the near field, the antenna can produce either an electric field or a magnetic field, but not an electromagnetic field. Because of this, the NFC receiver has to be within the transmitters' near field. So NFC communicates either by a modulated electric field or a modulated magnetic field. However, the communication cannot happen via radio. An NFC-enabled card can be considered as a secure smart card. It is therefore used for secure authentication and other applications where secure authentication is required; some of the examples are: access control, tap and pay, mobile payment/wallet, ticketing systems, hotel keys, and public transport passes.

The NFC typically uses the secure version of the I^2C interface for integration on the system. A typical connection will look like Figure 6.18.

In Figure 6.18, the RESET pin is used by host processor to reset the NFC chip, and the INT pin is used by the NFC chip to ask for the host's attention via interrupt. The data communication happens via I^2C (CLK and data pins). Since I^2C can have many devices on the bus, the NFC chip needs to have a unique identifier (with respect to the devices on that I^2C bus), known as a *slave ID*.

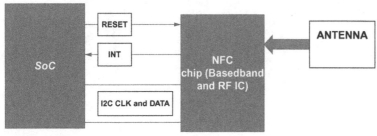

■ **FIGURE 6.18** NFC connection to a host processor/SoC over I^2C.

I²C

As already discussed in Chapter 5, the I²C (Inter-Integrated Circuit) bus is used for connecting low-speed peripherals to an SoC. It makes use of two lines for data transfer—a Data line (SDA) and a Clock line (SCL). Data is transferred serially over the I²C bus. Every I²C bus configuration can have

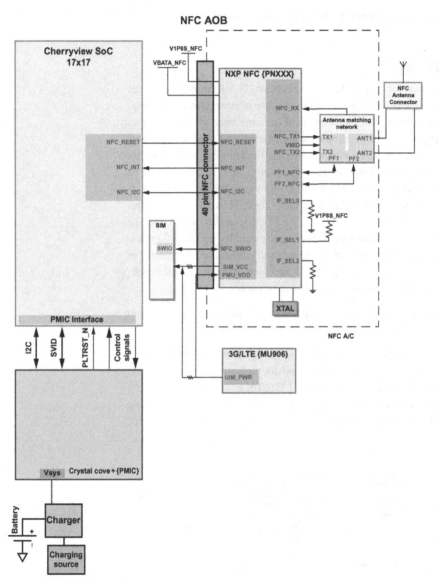

■ **FIGURE 6.19** A real world example of NFC integration.

one or multiple masters and one or multiple slaves. Each slave has an I^2C address that is used by the master to communicate with that particular slave.

I^2C is the preferred interface for integrating an NFC controller on the system because it's simple, well known, universally accepted, low-power, and cost effective. The only problem with I^2C is it is slow speed, which is not really a limiter for the kind of applications NFC is used for: NFC is typically used for authentication purposes, and the amount of data used for the purpose is small, and therefore the data transfer rate does not matter as much. Figure 6.19 shows a real-world example of NFC integration on a system via I^2C interface. The NFC module in use is from NXP semiconductors that are integrated to Intel's Cherryview SoC.

SUMMARY

In this chapter we quickly glanced over the most popular communication ICs on a mobile device. We discussed their functionalities in brief, since a detailed description is beyond the scope of this book. After discussing the functionalities in brief, we also discussed the various interfaces used for integrating the communication ICs and did a high-level comparison of the same with respect to the applicability in various usage scenarios.

7

Memory Interfaces

Following up on the general discussion of various interfaces in Chapter 3, this chapter discusses, in detail, the various interfaces for volatile and nonvolatile storage, their applicability to various scenarios, and their capabilities. After reading this chapter, the reader should understand when to use a particular interface and what advantages it has over other interfaces for a particular design.

VOLATILE MEMORY INTERFACE

There are two types of memory: volatile and nonvolatile. Typically the main memory (also known as the system memory) is volatile. The system memory is an important component in any system. The code and data that the CPU executes and operates respectively are stored in the system memory. In addition to this it is also used for storing the data that is operated on by other subsystems within the system on chip (SoC), like video or image processing. Because of this, the interface to the system memory should be robust enough to support high-speed transfer rates, yet be power- and cost-efficient. In this section we will see these aspects of the system memory interface.

System memory

Before understanding the system memory interface it is important to understand what type of memory is best suited for system memory. It is obvious that one would select random access memory (RAM) as the choice for system memory, because it is required to access the memory in a random fashion. As explained in Chapter 3, there are two types of RAM: static RAM (SRAM) and dynamic RAM (DRAM). DRAM uses one capacitor and one transistor to implement one bit of storage, whereas SRAM uses six transistors (basically 1 D flip flop) to implement one bit of storage. In order to save cost and power, DRAM chips are used as system memory. They give better memory density for the same number of transistors used. Since the bit is stored as a charge in the capacitor, which is prone to leak over a period of time, it is required to periodically refresh. This is an additional overhead for DRAM.

DRAM operation

DRAMs are organized as a matrix in the form of rows and columns as shown in Figure 7.1. One can consider the intersection of a row and column to be a memory cell. To access one memory cell, first the row address is provided and using this, the appropriate row is selected. This means that all the memory cells within the selected row can be accessed subsequently. This operation is commonly referred to as the *activate command*. Once this is done, the column address is provided to access the actual memory address of interest, which can be finally read or written. If it is a write operation, then the appropriate bit from the data buffer is written to the memory, and if it is a read operation, the selected memory bit is transferred back to the data buffer.

Figure 7.1 shows how a DRAM is organized as memory cells internally. One such matrix of the memory cells is referred to as a *bank* or *page*. And many such banks are present in the DRAM memory. Typically the size of a bank is 4 KB, or 8 KB, or the like. So as part of the activate command it is also required to specify the bank address. Typically the number of address bits used for bank is 2-3, for columns and rows it is in the range of 10-12 each. The address generated by an SoC is normally 36 bits; normally the lower bits

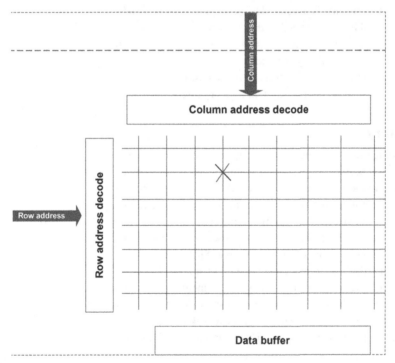

■ **FIGURE 7.1** DRAM memory.

are used for selecting the columns, the upper bits for rows, and then the banks. There can be other ways of deriving the bank, column, and row address from the SoC address bus. In order to save the pin count in the DRAM, the same address bits are used for both the column and rows. Basically they are multiplexed on the same set of address lines. Special control signals called row address strobe (RAS) and column address strobe (CAS) are used for differentiating the activate command from a read/write operation. If the subsequent access falls in the same bank and row as the previous access and a different column, then it is sufficient to just provide the CAS and the column address. In this case it is not required to activate the bank again. Normally such accesses are faster. However if it is required to select a different bank or row, then the current bank and row is closed; this operation is referred to as a *precharge*. Following the precharge a new activate command is followed along with RAS, following which the CAS can be provided. If the subsequent access falls in different banks or rows then memory latency will be higher. Table 7.1 summarizes some of the important timing parameters related to DRAM.

The parameters in Table 7.1 are commonly written as four numbers separated with dashes on the DRAM modules, such as 5-7 to 7-15. The CAS latency comes into picture while accessing an already activated row. It is the time it takes for the data to be available for a read operation after asserting the CAS. The RAS to CAS delay comes into play when a new row is being activated. The CAS can be asserted only after the activate command is successfully completed, hence the CAS cannot be asserted immediately after the RAS.

Apart from this the DRAM is also required to be refreshed periodically to prevent the loss of data due to leakage of charge from capacitors. A refresh

Table 7.1 DRAM Timing Parameters	
Timing Parameter	**Description**
tCL	*CAS latency.* This signifies the latency for accessing a column in a selected row
tRCD	*RAS to CAS delay.* This signifies the delay between an activate command (RAS) and CAS
tRP	*Row precharge time.* This is the time required to close an open row
tRAS	This is the number of clock cycles required between a bank active command and the issuing of the precharge command

command is used for such purposes. The memory controller within the SoC takes care of converting the SoC address bus into the bank, row, and column addresses and issuing the appropriate commands to the memory. The memory controller also takes cares of meeting the timing requirements.

Types of DRAM

DRAM memory has been used as the system memory for most of the applications. Over a period of time different types of DRAMs have evolved mainly to reduce the latencies of accesses, increase the size, and reduce power dissipation. The following section discusses the various types briefly.

DDR SDRAM

DRAMs are commonly synchronous in design, hence they were also known as SDRAM or synchronous DRAM. This means that a memory clock is used between the memory controller and the actual DRAM and all operations are carried out using the same. Over a period of time enhancements in SDRAM technology led to the introduction of the double data rate (DDR) SDRAM, which means that the memory can transfer data both on the positive and negative edge of the clock. This is also commonly referred to as double-pumping the data. With a data width of 64 bits, DDR SDRAM can achieve a data rate of (memory clock frequency) $\times 2$ (double pumping) $\times 64$ (data bus width). Some of the standard DDR data rates are listed in Table 7.2.

DDR2 SDRAM

Further enhancements to the DDR technology lead to the advent of DDR2 and DDR3. The DDR2 technology allows the external bus or I/O clock to be operated at twice the speed of the DRAM internal memory clock. Most of the time CPU will access sequential data, hence the DRAM can prefetch data from subsequent locations assuming that the CPU will request for the same in the future. Once data is prefetched, using a faster bus or I/O clock will reduce

Table 7.2 Some Standard DDR Transfer Rates

DDR Type	Memory Clock (MHz)	I/O Bus Clock (MHz)	Peak Data Rate (MB/s)
DDR-200	100	100	1600
DDR-266	$133\frac{1}{3}$	$133\frac{1}{3}$	$2133\frac{1}{3}$
DDR-333	$166\frac{2}{3}$	$166\frac{2}{3}$	$2666\frac{2}{3}$
DDR-400	200	200	3200

Table 7.3 Some Standard DDR2 Transfer Rates

DDR2 Type	Memory Clock (MHz)	I/O Bus Clock (MHz)	Peak Data Rate (MB/s)
DDR2-400	100	200	3200
DDR2-533	133⅓	266⅔	4266⅔
DDR2-667	166⅔	333⅓	5333⅓
DDR2-800C	200	400	6400
DDR2-1066	266⅔	533⅓	8533⅓

the latency in sending the data to the memory controller. By operating the bus or I/O clock at twice the rate of the memory clock, a maximum of four 64-bit words can be prefetched. This is the main advantage of operating the bus clock faster than the memory clock. With a data width of 64 bits, DDR2 SDRAM can achieve a data rate of (memory clock frequency) × 2 (for bus clock being twice of memory clock) × 2 (double pumping) × 64 (data bus width). Some of the standard DDR2 data rates are listed in Table 7.3.

DDR3 SDRAM

The key difference between DDR2 and DDR3 technology is the external bus or I/O clock is operated at four times the speed of the DRAM internal memory clock. By operating the bus or I/O clock at four times the rate of the memory clock, a maximum of eight 64-bit words can be prefetched. With a data width of 64 bits, DDR3 SDRAM can achieve a data rate of (memory clock frequency) × 4 (for bus clock being four times of memory clock) × 2 (double pumping) × 64 (data bus width). Some of the standard DDR3 data rates are listed in Table 7.4.

Table 7.4 Some Standard DDR3 Transfer Rates

DDR3 Type	Memory Clock (MHz)	I/O Bus Clock (MHz)	Peak Data Rate (MB/s)
DDR3-800	100	400	6400
DDR3-1066	133⅓	533⅓	8533⅓
DDR3-1333	166⅔	666⅔	10,666⅔
DDR3-1600	200	800	12,800
DDR3-1866	233⅓	933⅓	14,933⅓

With each generation of the DDR, enhancements have been made to increase the capacity of the DRAMs and provide faster data rates. This is evident from Tables 7.2–7.4.

DDR2 interface details—a case study

This section describes the DDR2 interface specification as per the JEDEC specification.

DDR2 signal list

The DDR2 interface is comprised of the following signals, which are explained here briefly:

- *CKp*, *CKn*—This is an input differential signal and it is the bus clock. All commands and operations are carried out synchronously with this clock.
- *CKE—Clock enable input signal*. This is used for enabling certain commands that will be discussed later.
- *CS—Chip select input signal*. This is used for selecting the right set of DDR2 memory chips. Typically in order to realize the 64 bits of data, multiple DDR2 chips are used. The width of the individual chips are either 8 or 16, in which case 8 or 4 of them are required to create 64 bits of data. For example if 16-bit DDR2 chips are used then data bits 0-15 are connected to the first memory chip, data bits 16-31 are connected to the second memory chip, data bits 32-47 are connected to the third memory chip, and data bits 48-63 are connected to the fourth memory chip. All the memory operations are simultaneously carried out in the four chips. Typically all these chips are placed on a PCB known as a memory module and connected accordingly. Sometimes to increase the memory capacity on the memory module one more set of such chips can be mounted. In this case it is required to select one of the two such sets of chips. This CS signal is used for differentiating the same.
- *RAS, CAS, WE*—Collectively known as the command input signals. The different commands like activate, read, write, etc. are realized using different combinations of these signals.
- *BA0-BA2—Bank address inputs*. They are used for specifying the bank address during an activate command.
- *A-A15—Address input*. These are used for specifying the row or column address.
- *DQ0-DQ63—Bidirectional input/output data.*
- *DM0-DM7—Data mask input signal*. Used as a data mask for write operations.

Burst operations

The DDR2 can access a minimum of 64 bits of data at a time. One can consider the word length of DDR2 accesses to be of 64 bits. In order to support a constant supply of data to the CPU burst operations are supported where in even the subsequent 64-bit data words are accessed. When the CPU is accessing in sequential manner, the burst accesses help reduce the latency. The lengths of the bursts determine how many of these are accessed subsequently. For a burst length of four (BL4), four 64-bit words are accessed in succession. And for a burst length of eight (BL8), eight 64-bit words are accessed in succession.

Also note that for accessing anything that is less than 64 bits, the signals data masks are used, which allow only selective bytes to be written. And in case of reads, the unwanted bytes are discarded by the memory controller.

NONVOLATILE MEMORY INTERFACE

The nonvolatile memory is used for storing data permanently like flash or hard disk. Interfaces to such devices will be discussed later in this section.

ROM

The term ROM, meaning *read only memory*, was originally used to refer to memory that can be only read and not written. In these ROMs, the data is physically encoded in the circuit, so it can only be programmed during fabrication of the ROM chip. Original ROMs used combinational logic to embed content in the circuit. While this served the purpose of nonvolatility, there were significant other limitation/problems with this:

1. Since the content was integrated in the circuit itself, the memory had to be redesigned even if only the content changed. It was therefore not economical if the number of chips to be made were only a few.
2. Even for just a change in the content, the turnaround time in getting the chip ready was significant, because it had to go through the complete process starting from the design of the chip.
3. Since the ROM cannot be changed, it is not suitable for experimental purposes during research and development of a product.
4. Once the product is shipped with faulty ROM, the only way to fix the problem was to recall the units and then physically change the ROMs on all these units.

The above four challenges drove the evolution of technology in the domain. Subsequent developments have addressed these challenges one after another. PROM, invented in 1956, allowed users to program its contents exactly once by physically altering its structure with the application of high-voltage pulses. This addressed problems 1 and 2 above, since one can buy or order fresh unprogrammed PROM chips and program them with the desired contents at one's convenience. The 1971 invention of EPROM essentially solved problem 3, since EPROM can be repeatedly reset to its unprogrammed state by exposure to strong ultraviolet light. Once the EPROM chips were reset to the fresh, unprogrammed state, they could be reprogrammed with desired content. EEPROM, invented in 1983, went a significant step further and solved problem 4, since an EEPROM can be programmed in-place by a means of providing data from an external source like a serial port. Flash memory, invented at Toshiba in the mid-1980s and commercialized in the early 1990s, is a form of EEPROM that makes very efficient use of chip area and can be erased and reprogrammed thousands of times without damage. The most recent development is NAND flash, also invented at Toshiba. The NAND flash improve the latency of reprogramming so much that it is comparable to other nonvolatile storage alternatives. By offering throughput comparable to hard disks, and higher tolerance to physical shock, the NAND flash is becoming a preferred choice for nonvolatile storage. Serial Peripheral Interface (SPI) is the most common interface for connecting SoC with ROM/FLASH memory.

SPI

SPI is a simple synchronous serial interface that is used widely in the industry to connect devices to the main application processors in the modern day ASICs. In SPI the signaling occurs through a set of four wires: SERIAL DATA IN, SERIAL DATA OUT, CLOCK, and CS. An SPI device can be a master or a slave depending upon who is driving the clock. The SPI standard allows for one master and multiple slaves on the bus. Because of the number of pins, sometimes SPI is called a *four-wire* serial bus; SPI is also referred to as SSI (synchronous serial interface).

It is in many ways similar to the I^{2}S interface. SPI competes with several contemporary interfaces like SSP (serial SCSI (small computer system interface) protocol) and Microwire, each having only a small set of differences. SPI is the prevalent interface of flash memory, EEPROM, and so on, but other devices like general purpose input output (GPIO) expanders, latches, and so on are also widely available in the market. Catering to the needs of nonvolatile memory, SPI devices can also be daisy chained.

Features. The following are the important features of SPI:

- Full-duplex four-wire synchronous interface
- Master and slave operation

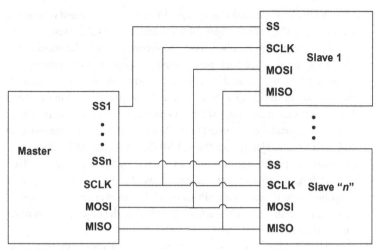

■ **FIGURE 7.2** A typical SPI system architecture.

- Option for daisy chaining
- Programmable clock polarity and phases

SPI system architecture. A typical SPI bus configuration will appear as shown in Figure 7.2.

As shown in Figure 7.2, the master connects with slave via a shared bus; however the CS signal is exclusive for each slave.

The four basic signals of SPI devices are denoted by SO (serial output) or MOSI (master out slave in), SI (serial input) or MISO (master in slave out), SCK (serial clock) or SCLK, and CS or SS (slave select), although various other similar nomenclature is quite common.

Each transaction begins with the master selecting a slave using the SS line. The master then proceeds to operate the serial clock (SCLK) at a frequency less than or equal to the maximum frequency supported by the slave (typical speeds are in the MHz range). In each clock cycle, the master sends one bit to the slave, and the slave sends one bit to the master. In the case where neither the master nor the slave has data to send, they can send 0s.

> *SO*—The SO line is configured as output in a master device and input in a slave device. They are also sometimes referred to as MOSI. Here the most significant bit is sent first. A programmable delay line is mostly present before the final presentation of the data by the master and just after the I/O in the slave. These delays are used to adjust the data latching with respect to the clock.

SI—The SO line is configured as input in a master device and output in a slave device. They are also sometime referred to as MISO. Here the most significant bit is sent first in a slave and received first in the master. The MISO line in a slave device is placed in high impedance state if the slave is not selected. A programmable delay line is mostly present before the final presentation of the data by the master and just after the I/O in the slave. These delays are used to adjust the data latching with respect to the clock.

SCK—The serial clock is used to synchronize the data transmission in and out of the device for MOSI and MISO lines. The SCK clock is generated by the master device and received by the slave device. The data latching polarity of both the master and the slave device are mostly programmable in almost all SPI devices. The master device always places data on the MOSI line a half cycle before the clock edge, in order for the data to latch.

CS—The CS line is use to select a slave device. To make the operation simple, generally two or three CS signals are present, depending upon the system requirement of the ASIC. The CS signal has to be low prior to the data transactions and must stay low for the duration of the transaction.

Functional description. The functionality of the SPI device is mostly based on the system requirement. For a SPI device that has to implement daisy chaining, both the MOSI and MISO signal carry the same data synchronous with the SCLK, although a definite delay of up to two clock cycles may be present in some implementations. For devices requiring full-duplex communication, MOSI and MISO may be tied to different data buffers. The data buffer is mostly of 8-bit width, and since there is only one CS signal between master and slave, the number of bits to be transmitted on either side must be known prior to the system program. This means either slave or master could have a situation in which it has nothing to transmit (indicated by low on the data line) but it continues to receive data synchronous with the master clock. This situation has been handled by the system program and the extra bit received needs to be padded.

Thus, the SPI consists of a buffer that interfaces with the system through DMA (direct memory access) or a CPU-addressable buffer. The clock logic derives its clock from the internal system clock and is programmable for the speed required. The pin control logic provides the output drives and the delay settings. There could be various control and status registers that would provide programmability to the device.

Multi I/O SPI device. A multi I/O SPI device is capable of supporting increased throughput from a single device. This is achieved by having more than one data lines. For example, a dual I/O SPI has 2-bit data bus interface and therefore it enables two times the transfer rate of the standard serial flash memory devices; similarly a quad I/O SPI has 4-bit data bus interface and therefore it is able to improve throughput by four times. Due to the increased bandwidth supported by the multi I/O SPI devices, they can be applicable to a much wider range of applications requiring higher performance.

MMC/SD

We already discussed the MultiMediaCard (MMC) and SD (secure digital) interface while discussing communication interfaces. We summarize that in this section and then discuss the storage-specific features of the SD interface here. The SD standard was introduced in August 1999 as an evolutionary improvement over MMC. The SD standard is maintained by the SD association. Effectively, SD is secure version of MMC, which enabled digital rights management for audio/video content. The SD format includes four card families available in three different form factors. The four families are the original standard-capacity (SDSC), the high-capacity (SDHC), the eXtended-capacity (SDXC), and the secure digital input output (SDIO), which combines input/output functions with data storage. The three form factors are the original size, the mini size, and the micro size. One can look at the web for pictorial illustration of the three form factors.

Physical dimension

The three form factors have different physical dimensions:

Standard size
- The respective families in standard sizes are SD (SDSC), SDHC, SDXC, and SDIO.
  - There are two variations of sizes: the regular size is $32.0 \times 24.0 \times 2.1$ mm ($1.260 \times 0.945 \times 0.083$ in.); however, there is another variation in standard size measuring $32.0 \times 24.0 \times 1.4$ mm ($1.260 \times 0.945 \times 0.055$ in.—as thin as MMC) for thin SD.

Mini size
- The respective families in mini size are miniSD, miniSDHC, and miniSDIO.
  - The size in mini variant is $21.5 \times 20.0 \times 1.4$ mm ($0.846 \times 0.787 \times 0.055$ in.).

Micro size

- The micro form factor is the smallest SD card format.
- The respective families in micro size are microSD, microSDHC, and microSDXC.
- The size in micro variant is $15.0 \times 11.0 \times 1.0$ mm $(0.591 \times 0.433 \times 0.039$ in.$)$.

Physical interface

From the interface standpoint, the SD interface has CLK (i.e., clock), CMD (i.e., command) and data signals.

The number of data pins are not fixed by the standard: it can use one or four data pins to support the speed design requirements.

The details of the SD bus and interface have been discussed in Chapter 6, "Communication Interfaces"; refer to that chapter for further details.

Pin diagram

The SD memory card exhibits nine pins (eight pins in the case of the microSD card) that are used to support electrical communication.

The SD bus uses the following signals:

- CLK: clock signal driven from host to card
- CMD: muxed command/response signal, bidirectional in nature
- DAT0-DAT3: four data signals, bidirectional in nature
- VDD, VSS1, VSS2: power and ground signals

In comparison to SD bus, the SPI bus uses the following signals:

- CS: chip select signal driven from host to card
- CLK: clock signal driven from host to card
- DataIn: data signal driven from host to card
- DataOut: data signal driven from card to host

The reason we discussed both SD bus pins/signals and SPI bus pins/signals in comparison in this context is for the fact that the SD controller can support both SD mode and SPI mode of operation on the same set of pins. The host system can choose either of the two modes. The card detects the mode requested by the host when the reset command is received and sets to operate in that mode; it therefore expects all further communication to be in the same mode. The mode of operation can be reset by a reset command initiated by the host.

Table 7.5 Pin Mapping in SPI and SD Mode

PIN	SD Mode		SPI Mode	
---	Name	Description	Name	Description
1	CD/DAT3	Card detect/data	CS	Chip select
2	CMD	Command/response	DI	Data in
3	GND1/VSS1	Ground	VSS	Ground
4	VDD	Supply voltage	VDD	Supply voltage
5	CLK	Clock	SCLK	Clock
6	GND2/VSS2	Ground	GND2/VSS2	Ground
7	DAT0	Data	DO	Data out
8	DAT1	Data	RSV	Reserved
9	DAT2	Data	RSV	Reserved

As evident from the preceding sections, SPI uses seven of the nine SD signals. DAT1 and DAT2 are not used; DAT3 is used as the CS signal. The pin mapping of various signals in SD and SPI mode are as indicated in Table 7.5.

SCSI

Before SCSI came to existence, typically there were device-specific interfaces used to communicate with specific devices. In other words, the interfaces used for interfacing a device varied with each device. For example, an hard disk drive (HDD) interface could only be used with a HDD. To alleviate the problem and allow more flexibility to the system designer, SCSI was developed to provide a device-independent mechanism for attaching to and accessing host computers. SCSI also provides an efficient peer-to-peer I/O bus that can support multiple devices. Even though today SCSI is primarily used as a hard disk interface, SCSI can be effectively used to interface other devices, such as tape drives and optical media drives, to the host computer. The changes required in the system to support these devices are minimal to nil.

What is SCSI?

The Small Computer System Interface (SCSI) is both a command set and a transport. It defines an interface and command set for interconnecting computers and peripheral devices.

SCSI command sets have been defined for many different device types and are carried on many different transports. The parallel SCSI transport currently is used primarily in file servers and small to mid-range disk arrays. It is marketed to fit between fibre channel at the high-end of storage arrays and Integrated Drive Electronics (IDE) or Advanced Technology Attachment (ATA) at the desktop or workstation end of the marketplace.

Serial attached SCSI (SAS) is a medium cost, high availability replacement for parallel SCSI, using a point-to-point connection for devices including initiators, targets, and expanders, and a routable protocol to connect over 16,000 end devices. SAS allows lower cost devices by providing connectivity for SATA (serial ATA) 2.x target devices.

SCSI history

Shugart Associates and NCR developed a system interface named SASI (short form for Shugart Associates System Interface). The interface was developed in 1981 and its primary objective was to build a proprietary, high-performance standard for use by the two companies. However, they realized the need to increase the acceptance of SASI in the industry, and in order to do that the standard was revised and published as SCSI. SCSI was formally acknowledged as an industry standard by ANSI (American National Standards Institution) in 1986. SCSI, first developed for hard disks, is often compared to IDE/ATA. Even though SCSI offers improved performance and expandability compared to other storage interfaces, which makes it suitable for high-end computers, the high cost factor associated with SCSI hinders its prominence among home or business desktop users. SCSI command sets were defined keeping disk drives, tape drives, printers, processors, and write once read many drives in mind; and over the years, SCSI has undergone significant changes to evolve into a prominent and viable industry standard for various devices. There has been multiple versions of SCSI standards and a brief overview of them will follow here.

SCSI-1 is the original standard approved by ANSI. It was renamed to distinguish it from other SCSI versions. SCSI-1 acts as the base and it defines the basics of the SCSI bus. The definition, includes cable length, signaling characteristics, commands, and transfer modes. SCSI-1 uses 8-bit bus thus limiting the maximum data transfer rate to 5 MB/s. In addition, SCSI-1 devices support only single-ended transmission and passive termination.

Since the SCSI-1 standard did not limit the command set that the implementers implemented, there were multiple implementations of SCSI-1 with differing command set. To contain the problems caused due to this, a working paper was created to define a set of standard commands for a SCSI device. This set of standards was termed as common command set (CCS). CCS formed the basis of the SCSI-2 standard. The key focus of SCSI-2 was to enhance reliability and interoperability by standardizing and formalizing

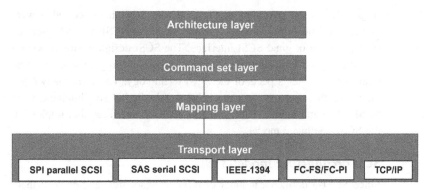

■ **FIGURE 7.3** SCSI layers.

the SCSI commands. Additionally it wanted improved performance over SCSI-1. Development of SCSI-3, the next version of SCSI standard, was started in 1993, Starting SCSI-3, a modular approach was taken while defining the standard. Therefore the SCSI-3 standard comprises of multiple related standards instead of being one monolithic specification document.

SCSI-3 broke SCSI protocol into four layers (see Figure 7.3). The top layer (architecture layer), standardized in a document called SCSI architectural model (SAM), defines what SCSI is. The next layer (command set layer) is a series of documents for command sets; there is roughly one document per device type. The third layer, called the "protocol" layer, is actually a mapping standard, which documents how to place the commands on the transport. The fourth layer (transport layer) standardizes the means of moving commands, data, and status from one device to another.

SCSI-1 and SCSI-2 each defined their command set and transport in a single document. To upgrade a command set (e.g., the tape drive), the entire document must be revised and approved by many organizations that may have no interest in tape drives. In SCSI-3, a single command set may be upgraded or a single transport may be added. Other aspects of SCSI-3 are not affected because each command set and each transport in SCSI-3 is its own document. In fact, nearly all the command sets and SCSI transports have been upgraded, some many times.

New device types and transports have been added under SCSI-3. With the flexibility of SCSI-3, it is not necessary to go to SCSI-4. SCSI-3 has added the following command sets: SCSI controller commands (disk arrays), enclosure services, reduced block commands, optical card reader/writer (OCRW), bridge controllers and object-based storage device. The "-3" was dropped from the name and now it is referred to as SCSI.

In parallel to the evolving SCSI standards, SCSI interfaces also went through several improvements. parallel SCSI, or SCSI parallel interface (SPI), was the original SCSI interface. The SCSI design is now making a transition into SAS, which is based on a serial point-to-point design, while retaining the other aspects of the SCSI technology. There are many other interfaces that are not complete SCSI standards, but still implement the SCSI command model. Table 7.6 lists the key interfaces that implement the SCSI command model.

SCSI architecture model-3

Since SCSI-3 defines a number of different standards, and each of them cover specific and different aspects of SCSI; it was necessary to organize them in a way to define their relation to each other, and the goals of the interface as a whole. This structure is called the *architecture* of SCSI-3. So, in other words, the SCSI-3 architecture defines and categorizes various SCSI-3 standards and requirements for SCSI-3 implementations. Details on SCSI architecture and requirement are documented in Technical Committee T10 "SCSI Architecture Model-3 (SAM-3)" document available at www. t10.org. The SCSI-3 architecture was approved and published as the standard X.3.270-1996 by the ANSI. The SCSI-3 architecture helps to understand and effectively utilize SCSI. There are three major components of the SCSI-3 architecture model: SCSI command protocol, transport layer

Table 7.6 Key Interfaces Implementing SCSI Command Model

Interface	Standard	Width	Clock	MAX Throughput (MB/s)	MAX Devices
SCSI-1	SCSI-1 (1986)	8	5 MHz	5	8
Fast SCSI	SCSI-2 (1994)	8	10 MHz	10	8
Fast-Wide SCSI	SCSI-2; SCSI-3 SPI (1996)	16	10 MHz	20	16
Ultra SCSI	SCSI-3 SPI	8	20 MHz	20	8
Ultra Wide CSI	SCSI-3 SPI	16	20 MHz	40	16
Ultra2 SCSI	SCSI-3 SPI-2 (1997)	8	40 MHz	40	8
Ultra2 Wide SCSI	SCSI-3 SPI-2	16	40 MHz	80	16
Ultra3 SCSI	SCSI-3 SPI-3 (1999)	16	40 MHz DDR	160	16
Ultra-320 SCSI	SCSI-3(2002)	16	80 MHz DDR	320	16
Ultra-640 SCSI	SCSI-3(2003)	16	160 MHz DDR	640	16

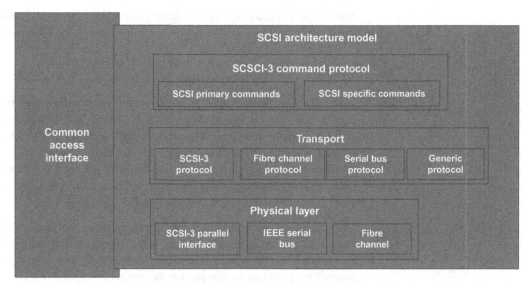

■ **FIGURE 7.4** SCSI architecture model.

protocols, and physical layer interconnects. Figure 7.4 shows the SCSI-3 standards architecture along with other standards within SCSI-3. In the following section we will briefly talk about the three components.

> *SCSI command protocols.* This consists of primary commands that are common to all devices. It also consists of device-specific commands that are unique to a given class of devices.
> *Transport layer protocols.* Transport layer protocols are a standard set of rules for devices to communicate and share information.
> *Physical layer interconnects.* Physical layer interconnects define the interface details such as electrical signaling methods and data transfer modes.

Initiators and targets

All communication in SCSI is between a single initiator (computer or computer-like device that can "initiate" commands) and a single target (peripheral device that executes those commands).

SCSI-3 architecture leverages or employs the client-server model of communication. As it is known for client-server model, a client initiates a service request to the server and the server fulfills the client's request. In SCSI environment, the initiator kind of acts as the client and target acts as the server. In a SCSI-3 client-server model, a particular SCSI device acts as a SCSI target device, a SCSI initiator device, or a SCSI target/initiator device. A device at a point can play one of the two roles while participating in an IO operation:

- *SCSI initiator*—Initiates a request to SCSI target device
- *SCSI target*—Services the request received from a SCSI initiator.

Most SCSI devices are used solely as an initiator or solely as a target. However, there are some devices that can switch modes and perform the duties of either an initiator or a target. Therefore, it may be more exact to refer to devices as operating in "initiator mode" or in "target mode." The terms "initiator mode" and "target mode" are normally not used except when speaking of a device capable of switching modes. For that reason, this text uses the simpler terms of initiator and target.

Each device service request contains a command descriptor block (CDB). The CDB opcode is an 8-bit structure. The CDB defines the command to be executed and it defines command-specific inputs and other control parameters specifying how to process the command.

SCSI ports are the physical connectors; and the SCSI cable plugs into ports for communication with a SCSI device. A SCSI device may contain target ports, initiator ports, target/initiator ports, or a target with multiple ports depending upon the nature and the purpose of the device. Depending upon the port combinations, a SCSI device can be classified as an initiator model, a target model, a combined model, or a target model with multiple ports. Since in an initiator model the SCSI device has only initiator ports, the application client can only initiate requests and receive confirmation and cannot serve any requests. Similarly, a SCSI device with only a target port can serve requests but cannot initiate them. The SCSI target/initiator device can switch role it plays while participating in an I/O operation. A SCSI device may also have multiple ports of the same orientation to support service requests from multiple devices

Each SCSI device (initiator or target) on a parallel SCSI bus is identified by a SCSI ID, a number between 0 and 7 (inclusive) on a narrow bus (1 byte wide) or between 0 and 15 (inclusive) on a wide bus (2 bytes wide). There may be any number of initiators and any number of targets, but the sum must not be greater than 8 or 16 and there must be at least one of each. These ID numbers set the device priorities on the SCSI bus. In narrow SCSI, 7 has the highest priority and 0 has the lowest priority. In wide SCSI, the device IDs from 8 to 15 have the highest priority, but the entire sequence of wide SCSI IDs has lower priority than narrow SCSI IDs. Therefore, the overall priority sequence for a wide SCSI is 7, 6, 5, 4, 3, 2, 1, 0, 15, 14, 13, 12, 11, 10, 9, and 8. SCSI allows for automatic assignment of device IDs on the bus while device initialization is taking place. This prevents two or more devices having the same SCSI ID.

Since SCSI follows the client-server model in communication, there need to be identifiers for initiators and targets. In the communication, an initiator ID uniquely identifies the initiator and is used as the originating address of the request. A target ID uniquely identifies a target who should serve the request coming from initiator. The initiator and target ID both are in the range of 0-15.

SCSI commands are now carried on the following transports:

- Parallel SCSI
- SAS
- Fibre channel
- Serial storage architecture
- TCP/IP (Internet SCSI)
- Parallel IDE/serial ATA (ATAPI)
- Printer interface
- USB
- IEEE 1394

SCSI is a set a standards. The set of standards defines a reference model that specifies common behaviors for SCSI devices. In addition it also defines an abstract structure generically applicable to all SCSI I/O system implementations.

The set of SCSI standards specifies the interfaces, functions, and operations necessary to ensure interoperability between conforming SCSI implementations. This standard is a functional description. Conforming implementations may employ any design technique that does not violate interoperability. The following concepts from previous versions of this standard are made obsolete by this standard:

- Support for the SPI-5 SCSI transport protocol
- Contingent allegiance
- The TARGET RESET task management function
- Basic task management model
- Untagged tasks
- The linked command function

SCSI command model

In the SCSI communication model (regardless of interface type: Parallel SCSI, SAS, or FA_AL2), the initiator and the target communicate with each other using a command protocol standard. The original SCSI command architecture was originally defined for parallel SCSI busses and later adopted for iSCSI and serial SCSI. There are other technologies that use the SCSI command set. Some of the examples of these technologies are

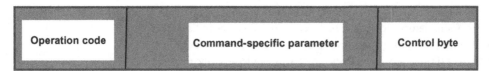

| Operation code | Command-specific parameter | Control byte |

■ **FIGURE 7.5** Command descriptor block (CDB).

ATA Packet Interface, USB Mass Storage class, and FireWire SBP-2. The SCSI command are defined using CDB (see Figure 7.5). CDB is "0" to "n" byte sequence where first byte is Operation code, "1" to "$n-1$" byte is command-specific parameter and byte "n" is control.

A CDB is command descriptor block and is accompanied by a list of parameters in the data-out buffer. The specific commands define the list of parameters valid for them. Table 7.7 lists the key commands in the SCSI command set.

Blocking and sharing

Because parallel SCSI is a shared bus, every device sees all signaling transmitted by any device. If one device is transferring data on the bus, then no other device may transfer data. This makes the bus architecture of parallel

Table 7.7 Common SCSI Commands

Command	Description
READ	Reads data
WRITE	Writes data
TEST UNIT READY	Queries the device for data transfer
INQUIRY	Returns basic information
REPORT LUNS	List the logical unit numbers
SEND AND RECEIVE DIAGNOSTIC RESULTS	Runs a self-test or a specialized test
FORMAT UNIT	Sets all sectors to all zeroes and allocates logical blocks
LOG SENSE	Returns current information from log pages
LOG SELECT	Used to modify data in the log pages
MODE SENSE	Returns current device parameters from mode pages
MODE SELECT	Sets device parameters on a mode page

SCSI a "blocking" architecture, greatly limiting the performance on more heavily loaded busses. The opposite of this is the nonblocking architecture of most SAS expanders. Each device on an SAS expander may be communicating with the expander at exactly the same time. Physical links (the wires) between an SAS device and its expander are dedicated to that SAS device and expander and are thus not shared.

On a parallel SCSI bus with 16 devices, only one pair may communicate at a time. On a typical SAS expander with 16 devices, all 8 pairs may communicate at the same time. Note that the SAS standard requires only one communication link within the expander. All commercial expanders currently have many more than one internal link.

Problems with parallel SCSI

Wide parallel SCSI has 16 data lines, 2 parity lines, and 2 clocks for a total of 20 lines, all switching at high speed, causing a lot of radiated energy. The FCC has placed restrictions on how much of that energy may be radiated, requiring system manufacturers and system integrators to provide adequate shielding. With so many lines so close together, the signals of one line degrade the signals of the neighboring lines through inductive and capacitive coupling (crosstalk). Each of the twenty lines requires hardware driver circuitry. With 20 drivers switching simultaneously, there are huge power fluctuations, requiring chips to be designed for greater power dissipation. Drivers in chips require large space, use a lot of power and create a lot of heat; tolerances must be allowed for these problems and designs must take these problems into account.

Each wide bus may have between 2 and 16 (inclusive) devices on it. The drivers, cables, receivers and terminators must all be designed to work with 2 loads, 16 loads or any number of loads between 2 and 16 loads. Each of the drivers, each of the cable wires and each of the receivers, works at slightly different speeds (hopefully all within tolerance). When those differences are added together, some of the bits may arrive at the output of the receivers earlier than other.

The point-to-point serial solution

One answer to the problems of parallel SCSI is to use a point-to-point serial transport. The parallel information on the left is serialized and sent one bit at a time across the interface to the receiver, which deserializes the bits back into a parallel format. The chip that performs this function is called a SER/DES (short for serializer/deserializer).

Benefits of point-to-point serial transport

Following are the key benefits of a point-to-point serial interface:

- Eliminates the problems associated with a multi-drop bus.
- Knows how many devices are connected—it is always exactly two
- Allows each device to have exclusive access to the media as opposed to the shared bandwidth bus of parallel SCSI
- Requires only one driver in the chip
- Eliminates skew
- Eliminates crosstalk on the cable
- Reduces radiation
- Uses smaller and more flexible cable
- Uses smaller connectors

The new cables allow smaller enclosures with better air flow. The smaller connectors allow smaller disk drives, tape drives or other targets.

Other existing serial interfaces

There are many serial interfaces in use today that could be used for storage, some of them already using the SCSI command set. Why, then, should we invent a new transport?

Fibre channel

Fibre channel has been the high-end interface, architecturally able to connect nearly 16 million devices at distances measured in kilometers between devices. Fibre channel has been designed to transport many command sets, includes many features and many operating modes—all of which add to complexity and cost. The costs of fibre channel switches and end devices are very high. Fibre channel was started around 1988 and progressed slowly and carefully. The predominant clock rate today is 4.250 Gb/s, giving a byte data rate of 400 MB/s. To be fair in comparing this to parallel SCSI at 320 MB/s, the fibre channel transport is nonblocking (in fabric mode) but the parallel SCSI uses a blocking transport. Fibre channel products are also available at 8 and 10 Gb/s.

Serial ATA

SATA was designed for use with a single drive in a single computer. Since the original design, port multipliers have been designed to add more devices to a SATA connection and port selectors have been designed to allow more than one host to have access to a SATA drive. Even with these new devices, the SATA infrastructure is not adequate for high-end disk arrays. Port multipliers provide access to more drives, but the bandwidth is shared.

The ATA command set was designed for desktop applications and thus is not adequate for file servers. The information defining failures provided by ATA consists of one 8-bit register. The sense data provided in response to the SCSI request sense command is at least 18 bytes long.

SATA drives are designed for start-stop operations; whereas SCSI drives are designed for continuous operation. The "middleware7" programs used by many enterprises are based on the SCSI command set. To convert those programs to the ATA command set requires rewriting many millions of lines of code, a task the manufacturers are not willing to do, and for which customers are not willing to pay. There is a committee operating within the ATA and SCSI committees of ANSI to develop an open-system translation between ATA and SCSI commands.

ATA: ATA is a set of standards. Figure 7.6 shows the relationship between various ATA standards related to the ATA and SCSI families of standards and specifications. The set of Advanced Technology (AT) attachment standards consists of the AT attachment—8 ATA/ATAPI Architecture Model (ATA8-AAM), ATA-8-ACS, ATA8-AST, and other related host standards. The AT Attachment ATA Command Set (ATA8-ACS) specifies the command set host systems use to access storage devices. It provides a CCS for various parties involved in the system development and deployment, for example: systems manufacturers, system integrators, software developers, and suppliers of storage devices.

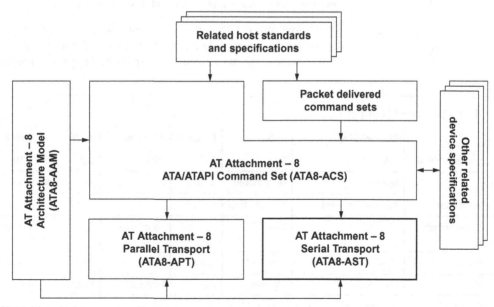

■ **FIGURE 7.6** Relationship between various ATA standards.

ATA protocols. ATA protocols are described as part of ATA transport documents. There are protocols that have to be implemented by all transports that use ATA8-ACS commands. The following set of protocols is described in ATA8-AAM:

- Non-data command protocol
- Programmed IO(PIO) data-in command protocol
- PIO data-out command protocol
- DMA command protocol
- PACKET command protocol
- DMA queued command protocol
- Execute device diagnostic command protocol, and
- Device reset command protocol.

The details and the implementation of each protocol is described in the transport standards and out of scope of this book.

USB

USB was originally designed for low and medium speed devices such as mice, keyboards, and still cameras. Later, high speed was added in USB 2.0 bringing the speed up to 60 MB/s, but on a shared bus. This is still too slow for mainstream disk applications although it serves very well for use with video, audio, and backups of desktop computers.

USB has recently defined super-speed USB which clocks at 5 Gb/s, giving a theoretical maximum speed of 500 MB/s. Table 7.8 gives a quick comparison of some of the more common transports in the storage industry.

Table 7.8 Comparison of Transports

	Parallel SCSI	SAS	FC-Loop	FC-Fabric	SATA	IEEE-1394
Maximum speed (MB/s)	320	600	200	1000	600	100
Transport	Shared	Non-shared	Shared	Non-shared	Non-shared	Shared
Maximum distance	12 m end to end	8 m per hop	[a]	[a]	1 m	[a]
Maximum number of devices	16	16 K	126	16 million	2	63 per bus
Command set	SCSI	SCSI	Many	Many	ATA	Many
Cost	Medium	Medium	High	Highest	Lowest	Low

[a]There are many different copper and optical transports defined, each with a different length.

SAS

SAS is an evolution of the traditional SCSI interface and it incorporates the decades of technology development in the domain. Work on the SAS specification was started by Compaq/HP, LSI Logic, Maxtor, and Seagate Technologies. They set upon the task of writing a specification in 2001.

The goal of the specification was to describe a new physical interface providing a roadmap for future performance improvements, while maintaining the elements of SCSI that worked well. The group wanted to keep the best features of the parallel SCSI bus, such as performance, reliability, availability and data integrity, while adding new features.

The group working on SAS agreed to leverage as much as possible from previous efforts (for both SCSI and other storage interfaces) and take advantage of the latest technologies available.

The SAS specification leveraged the mechanical component designs that were developed for SATA. By using similar, compatible connectors and cables, it would also be possible to use both SATA and SAS devices in the same system if the physical layer of the SAS interface was designed to be compatible with the physical layer of the SATA interface. This would provide customers with maximum flexibility in system configuration.

The initial SAS specification was targeted to have a 3-Gb/s data rate, twice the speed of SATA, since the SAS specification was based on a more advanced transceiver technology.

The SAS serial interface was also defined as full-duplex as opposed to half-duplex for SATA.

Another of the key goals for the SAS effort was to maintain or expand the maximum number of devices allowed in parallel SCSI domain. The parallel SCSI domain allowed for a maximum of 16 that could coexist. A SAS system with expanders may address up to 16,256 devices in a single SAS domain.

Leveraging elements of the SCSI protocol, including fibre channel where appropriate, was another of the initial goals for SAS, because there was a strong desire on the part of all of the members of the working group to provide maximum compatibility with current infrastructure and minimize the risk and cost of transition to the new interface.

Point-to-point configurations provide for high bandwidth. However, it requires intermediary devices between initiator devices (or hosts) and target devices (or peripheral devices) to support a topology where there may be more than two devices in a system. Inexpensive expanders act as the intermediary devices defined for SAS.

SAS expanders allow for systems in which more than one initiator may have a connection to more than one target (as is allowed by parallel SCSI).

SAS expanders are also introduced so that end devices connecting to expanders may be either SAS devices or SATA devices. It satisfies the goal of allowing heterogeneous system configurations.

The SAS architecture utilizing full-duplex point-to-point and wide ports provides high-performance capability. SAS expanders are high-speed switches that enable simultaneous communication between multiple initiators and targets.

The SAS architecture enables storage system with high reliability and availability by using redundant controllers, nonblocking expanders, and dual-port SAS disks. Dual ports enable SAS devices to be connected to multiple hosts; should one controller fail, the SAS device automatically switches to another available controller.

A wide port is created when there is more than one physical links in the port. A narrow port is created when there is only one physical links in the port. Wide ports significantly improve throughput by enabling several disks to communicate with a single port address at the same time. For example, a SAS controller with four 3-Gb/s links configured as a wide port will support data transfer rates over 12 Gb/s.

Currently, a 15 K-RPM disk drive will sustain data rates up to 75 MB/s. At these sustained data rates, a mere five disk drives will saturate an Ultra320 SCSI bus or a Fibre Channel arbitrated loop running at 2 Gb/s.

For SAS devices, the protocol highly leverages protocols described in other SCSI standards (such as the SCSI parallel interface, or SPI, standard), which are based on the SCSI-3 standards.

There are two methods to connect hard drives to host: direct attach and expander attach. In the direct attach the number of drives connected to the system are same as number of ports on host. In the expander attach, the drives are connected via expander and therefore the number of drives connected can be more than the number of ports on the host.

There are three protocols within SAS. Only a brief description is provided here.

- *SSP (serial SCSI protocol)*: It defines how the SAS transport transfers SCSI commands, data, responses (status), and control between the initiator and the target; following initialization and configuration, SSP should be the most frequently used protocol.

- *STP (SATA tunneled protocol)*: It defines how the SAS transport transfers ATA commands, data, status, and control between the SATA enabled SAS host and the SATA device. The SATA device in this case does not know of the SAS environment; the translation is done entirely in the SAS expander.
- *SMP (serial management protocol)*: It defines how an initiator can configure and maintain the expanders and SAS service delivery subsystem.

Transfer of commands, data, status, and other information for SAS devices is accomplished using packets very similar to the packets defined for fibre channel.

The format of SAS packets (called *frames*) is almost identical to fibre channel packets. The payload of the packets consists of CDBs and other SCSI constructs defined in other SCSI standards (such as the SCSI primary command set and SCSI block commands).

SAS protocol layers. The SAS protocol is divided into four layers: the phy layer, the link layer, the port layer and the transport layer. These four layers are contained in the SAS port. This means that applications (like software and drivers) used to communicate with parallel SCSI ports may also be used to communicate with SAS ports with little or no modification. In the section below we briefly talk about the various protocol layers.

> *Physical layer*—The physical layer is the cables, connectors, and so on.
> *Phy layer*—The phy layer is the electronics that drives the bits onto the cables or receives the bits from the cables. It also includes the OOB, 8b/10b encoding and decoding, SER/DES (serializer/deserializer), and clock recovery.
> *Link layer*—The link layer places the start of frame, cyclic redundancy check (CRC), and end of frame around the frame contents on transmit and removes them on receive. It checks the CRC, performs frame level flow control, and generates or handles primitives.
> *Port layer*—The port layer handles connections and disconnections. This is the layer that selects which phy of the possible phys available in a wide port will be used.
> *Transport layer*—The transport layer is responsible for constructing and parsing frame contents. The SSP transport layer handles SCSI commands, task management functions, data, and responses. The transport layer only receives frames for which the link layer will send an ACK. The SMP transport layer handles management functions for the

service delivery subsystem. Basically, it controls and configures the expanders. The STP transport layer handles ATA commands.

Application layer—In the SSP protocol, the application layer defines the mapping of SCSI commands, tasks and other SAM defined services to the SAS Transport Protocol Services. In the SMP protocol, the application layer defines the request and response frames used to manage the expanders of the service delivery subsystem.

The developments in the storage standards and interfaces are driven by four primary requirements of a storage device: high bandwidth, high capacity, low power, and low cost. Flash storage technology and standards have evolved rapidly to meet these requirements. Universal flash storage (UFS) was created with these four primary requirements in mind. In the next section we discuss UFS and its architecture.

UFS

UFS is a storage specification for flash devices. It is the next generation flash storage that provides the low power of eMMC with the high performance of SCSI solid state drive. It aims to provide a universal storage interface for both embedded and removable flash memory based storage in mobile devices such as smart phones and tablet computers. The specification is defined by JEDEC Solid State Technology Association: JEDEC Standard JEDS220.

The standard leverages existing standards such as the SCSI command set, the MIPI Alliance M-PHY, and UniPro. It also reuses the eMMC form factors to simplify adoption and development. UFS uses SCSI for upper layer, and M-PHY/UniPro for lower layer. UFS uses MIPI M-PHY as the physical layer and MIPI Unipro as the link layer. By supporting multiple gears as in M-PHY, the UFS provides scalable data rate with respectively different power consumption.

UFS system architecture

The UFS system architecture is illustrated in Figure 7.7.

The CPU talks to the UFS device via the USF host controller interface. As can be seen in Figure 7.7, the interconnect bus uses M-PHY and Unipro, while the UFS device leverages the SCSI command set and operations for implementation.

SCSI has already been discussed earlier in this chapter, while M-PHY/ Unipro has been discussed in Chapter 4, "Display Interfaces." Refer to those earlier discussions for details.

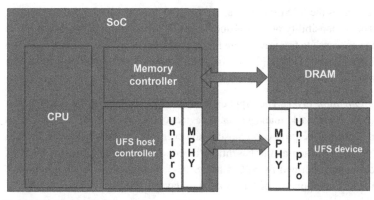

■ **FIGURE 7.7** High level system architecture with UFS.

UFS communication architecture

UFS has a layered communication architecture, which is based on SCSI SAM-5 architectural model. SAM-5 stands for SCSI architecture model-5 which defines, at abstract level, how SCSI devices should communicate. UFS communication architecture consists of the following layers: UCS (UFS Command Set), UTP (UFS transport protocol), and UIC (UFS interconnect), as shown in Figure 7.8.

In the following section, we will talk about the layered architecture of UFS system architecture: UCS, UTP, and interconnect (UniPro + M-PHY). The command set layer (UCS) is the interface to the software application. It

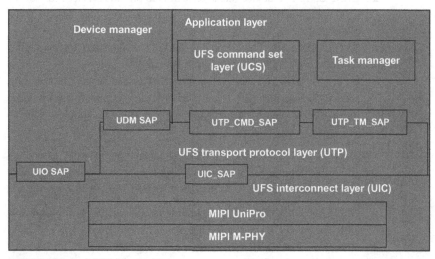

■ **FIGURE 7.8** Detailed UFS layered architecture.

leverages the SCSI standard as the baseline protocol for UFS specification. The responsibility of the transport layer (UTP) is to frame the protocol into the appropriate frame structure for transmission over interconnect layer. The interconnect layer (UIC) has digital and analog IP responsible for actual transfer of data.

Application layer. The Application layer has three primary components: UCS layer, task manager, and device manager. The UFS interface is designed to be protocol-agnostic; however, for versions 1.0 and 1.1, SCSI has been chosen as the baseline protocol. UFS supports a subset of SCSI commands defined by SPC-4 and SBC-3. Below is a brief description of the key components of the application layer:

- *UCS* handles SCSI commands supported by the UFS specification.
- *Task manager* handles task management functions defined by the UFS that are meant for command queue control.
- *Device manager* handles device level operations and device configuration operations. Device level operations constitute the device power management operations and commands to interconnect layers. Device level configurations constitute handling of query requests that are used to modify and retrieve configuration information of the device.

UTP layer. UTP layer is responsible for providing services to the higher layers via service access points (SAPs). As shown in Figure 7.8, UTP defines three SAPs for higher layers:

- UDM_SAP: The device manager SAP is to provide services for device level operations. These device level operations are done through query requests.
- UTP_CMD_SAP: The command SAP is responsible for providing services to UCS layer to transport commands.
- UTP_TM_SAP: The task management SAP is responsible for providing services to task manager to transport task management functions. UTP transports messages via UFS protocol information unit also known as UPIU.

UIC layer. UIC is the lowest layer of UFS layered architecture. It handles the connection between UFS host and UFS device. UIC consists of MIPI UniPro and MIPI M-PHY. UIC provides two SAPs to upper layers:

- UIC_SAP: UIC_SAP is responsible to provide services for transporting UPIU between UFS host and UFS device.
- UIO_SAP: UIO_SAP is responsible to provide services for issuing commands to UniPro layers.

It must be noted that the SCSI has been discussed in detail in one of the earlier sections of the chapter, and M-PHY has been discussed within the section on MIPI-DSI in Chapter 4, "Display Interfaces." Refer to these earlier sections for more details on SCSI and M-PHY as required.

CONCLUSION

In this chapter we quickly glanced over the memory interfaces for both volatile and nonvolatile types. We see that there are four primary drivers for evolution or advancement in memory technology and interfaces: high bandwidth, high capacity, low power, and low cost. The same drivers will continue to drive future technologies and interfaces in the time to come.

Chapter 8

Security Interfaces

This chapter describes the interfaces that are commonly used for secure elements such as TPM (Trusted Platform Module) and NFC (Near Field Communication). A few of the interfaces referenced in this chapter are already covered in detail in earlier chapters. In such cases, only the modifications made to the interface for added security are emphasized, and the reader is expected to refer back to the earlier chapters for basic details on the interface.

NEAR FIELD COMMUNICATION

NFC is a set of standards for portable devices to communicate with each other when they are brought into close contact, in the order of a few centimeters. NFC supports the following data rates: 106, 212, 424, and 848 kbps. NFC communication can be between two devices with NFC capability or between an NFC-capable device and a *tag*. A tag is nothing but an unpowered NFC chip.

NFC has a variety of use cases, a few of the more prominent ones being

- Contactless payment systems
- Mobile ticketing on public transportation
- Social networking—for sharing photos, and so on
- Bootstrapping more capable wireless connections

From the use cases mentioned here, it is obvious that the NFC communication needs to be secure, due to the very nature of the transactions.

Security threats

While designing a system that has an NFC interface, the designer must be aware that the following threats exist.

Eavesdropping

Eavesdropping is when a malicious device listens in on a valid NFC transaction. The malicious device can then use the information gathered to carry out spurious transactions on its own.

Data corruption and manipulation

Data corruption is when a malicious device interferes with the NFC data that is being transmitted to a reader so that the data, when it reaches the reader, is useless.

In data manipulation attacks, the malicious device modifies the data that is being transmitted to a reader.

Interception attack

This is similar to the man-in-the-middle attacks (see Figure 8.1). In this, a malicious device sets itself up between two genuine NFC devices and acts as the "other" device to both of them. It essentially acts as a middleman between the two devices and thereby gains access to all the data that is transmitted across.

Security provisions in NFC

One of the mitigating factors for security problems in an NFC configuration is that the NFC link operates only over short distances. So, for the man-in-the-middle type of security breaches, the attacker must be very close to the devices that are communicating. Apart from this, the NFC protocol itself does not have any provision for security built into it.

To prevent unauthorized devices from eavesdropping or participating in NFC communications, applications generally encrypt the data before sending it out. The encryption scheme used is flexible. Some example schemes are Advanced Encryption Standard (AES) and Data Encryption Standard (DES). Considering the keys for decrypting the data are available only with the authorized devices, the problem posed by malicious devices is negated with this measure.

There is one more source of security threat we have not discussed yet. For security-critical applications such as the ones mentioned earlier in this section, knowing that the data purportedly coming in from the NFC chip

■ **FIGURE 8.1** Interception attack—the communicating NFC devices think the malicious NFC device is the other device they are communicating with.

is in fact coming from the NFC chip and not from any malicious application is crucial. The Secure I²C interface ensures this and is the preferred interface for connecting NFC chips to an application processor or system on chip (SoC).

SUITABILITY OF I²C FOR NFC

Since NFC is a low-bandwidth interface in an embedded system, the NFC chip is typically connected to the application processor via a low-speed serial interface, usually I²C. I²C makes use of two lines for data transfer—a data line (serial data [SDA]) and a clock line (serial clock [SCL]). Data is transferred serially over the I²C bus, one bit of data per clock pulse. I²C supports clock speeds up to 5 MHz. Even after allowing for protocol overhead, this is more than adequate for supporting the 828 kbps maximum transfer rate required by NFC.

Secure I²C

In Secure I²C, there is an added layer of encryption on top of the I²C protocol. The encryption scheme used is flexible and can be chosen by the platform developer. Some example schemes are the two mentioned earlier, AES and DES.

The following sequence of steps explains the operation of Secure I²C protocol. The transaction is a write (from master to slave).

1. The master asserts a START condition, writes the slave device address, and clears the R/W bit. These operations follow the 7-bit addressing mode defined in the I²C standard.
2. Once the slave acknowledges the address, the master begins to send the data. This data is encrypted using AES, DES, or a similar encryption algorithm.
3. The master asserts the STOP condition.
4. The slave realizes that the communications have ended, and it can decrypt and check the signature, to verify the authenticity of the data received.

Read transactions (from the slave to the master) also follow the same principle. That is, the slave encrypts the data and sends it to the master. The master then decrypts the data after the transaction is completed.

Figure 8.2 shows the sequence of operations performed in realizing the Secure I²C implementation.

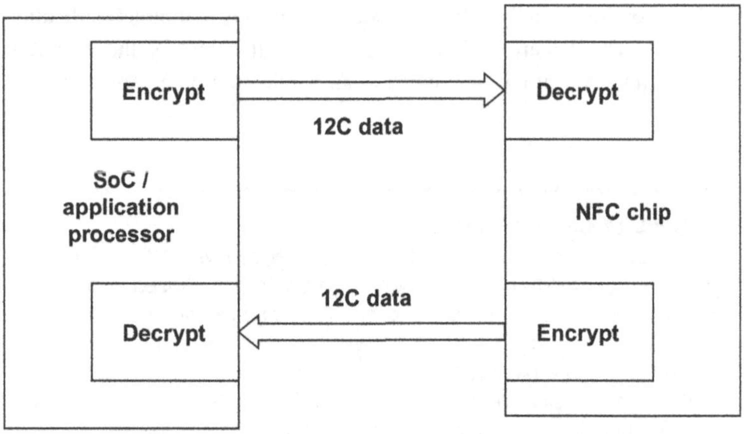

■ **FIGURE 8.2** Secure I²C implementation. Data received is matched against the pre-agreed encryption key and verified for authenticity.

To maintain the integrity of the Secure I²C protocol, the encryption keys used for communication must not be accessible to untrusted applications. This is typically taken care of by placing them in a protected memory region and allowing access only to some trusted drivers/applications.

TRUSTED PLATFORM MODULE

TPM is a core component in specifications developed by the Trusted Computing Group. These specifications describe how to add trust to existing devices and how to extend security in certain situations (e.g., in a network). The most common implementation of TPM is a small tamper-resistant chip capable of securely storing cryptographic keys and other security-critical information and providing cryptographic functions. The TPM chip does not initiate any transactions on its own. It only reacts to requests that are made to it.

The key difference between a TPM chip and the SoC is that the SoC should be easy to interact with and program, and by extension, easy to attack. The TPM chip is isolated behind a controlled interface and therefore is difficult to attack.

The Low Pin Count (LPC) interface and the Serial Peripheral Interface can be used for connecting TPM chips to an SoC or application processor. The LPC interface is typically used.

LPC interface

The LPC interface allows the legacy I/O motherboard components, typically integrated in a Super I/O chip,[1] to migrate from the ISA/X-bus to the LPC interface, while retaining full software compatibility. The LPC specification offers several key advantages over ISA/X-bus, such as reduced pin count, for easier, more cost-effective design.

Goals of the LPC interface

Here are the most important goals of the LPC interface:

- Enable a system without an ISA or X-bus.
- Intended for use by devices down on a motherboard only (no connector).
- Meet the data transfer rate of X-bus, and exceed those data rates where appropriate.
- Perform the same cycle types as the X-bus: memory, I/O, direct memory access (DMA), and bus master.
- *Synchronous design*: Much of the challenge of an X-bus design is meeting the different, and in some cases conflicting, ISA timings. Make the timings synchronous to a reference well known to component designers, such as Peripheral Component Interconnect (PCI).
- *Software transparency*: Do not require special drivers or configuration for this interface. The motherboard BIOS should be able to configure all devices at boot.
- Ability to support wake-up and other power state transitions.

Supported device classes

Only the following classes of devices can be connected to the LPC interface:

- Super I/O to I/O slave, DMA, bus master (for infra red (IR), parallel port (PP))
- Audio, including AC'97 style design to I/O slave, DMA, bus master
- Generic application memory, including BIOS to memory slave
- BIOS firmware memory to firmware memory slave
- Embedded controller (the TPM chip falls under this category) to I/O slave, bus master

Signal definition

There are seven required signals for the LPC interface. In addition, there are six optional signals that may be supported. Tables 8.1 and 8.2 list them all.

---

[1]A super I/O chip just combines interfaces for a variety of low-bandwidth devices. A typical super I/O chip will have a serial port, parallel port, infrared port, keyboard controller, and floppy disk controller.

Table 8.1 LPC: Required Signals

Signal	Peripheral (LPC Device)	LPC Host	Description
LAD [3:0]	I/O	I/O	Multiplexed command, address, and data
LFRAME#	I	O	*Frame*: Indicates start of a new cycle, termination of broken cycle
LRESET#	I	I	*Reset*: Same as PCI reset on the host. The host does not need this signal if it already has PCIRST# on its interface
LCLK	I	I	*Clock*: Same 33 MHz clock as PCI clock on the host. Same clock phase with typical PCI skew. The host does not need this signal if it already has PCICLK on its interface

Table 8.2 LPC: Optional Signals

Signal	Peripheral (LPC Device)	LPC Host	Description
LDRQ#	O	I	*Encoded DMA/bus master request*: Only needed by peripherals that need DMA or bus mastering. Requires an individual signal per peripheral. Peripherals may not share an LDRQ# signal
SERIRQ	I/O	I/O	*Serialized IRQ*: Only needed by peripherals that need interrupt support. This signal is required for the host if it does not contain the ISA IRQ lines as inputs
CLKRUN#	O	I/O	*Clock run*: Same as PCI CLKRUN#. Only needed by peripherals that need DMA or bus mastering in a system that can stop the PCI bus (generally in mobile systems) This signal is optional for the host
LPME#	O	I/O	*LPC power management event*: Similar to PCI PME#. Used by peripherals to request wake-up from a low power state
LPCPD#	I	O	*Power down*: Indicates that the peripheral should prepare for power to be removed from the LPC I/F devices. Actual power removal is system dependent. This signal is optional for the host
LSMI#	O	I	*SMI#*: Only needed if peripheral want to cause SMI# on an I/O instruction for retry. Otherwise can use SMI# via SERIRQ. This signal is optional for the host

From Tables 8.1 and 8.2 it can be seen that many of the signals are the same as signals found on the PCI interface and do not require any new pins on the host.

Block diagram of a system with an LPC interface

A typical block diagram for an LPC interface, showing the connection of a TPM controller to the application processor/SoC is shown in Figure 8.3.

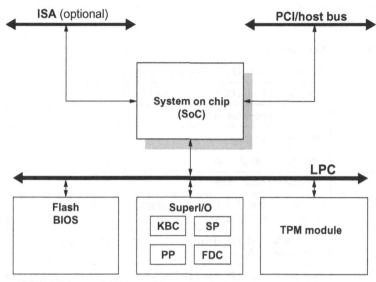

■ **FIGURE 8.3** An example LPC bus configuration.

Protocol overview

Data transfer in LPC occurs in cycles. Table 8.3 shows the cycles supported by the LPC protocol.

In general, a cycle proceeds like this:

1. A cycle is started by the host when it drives LFRAME# active and indicates a "start." For memory, I/O, and DMA cycles, the LAD [3:0] lines are set to [0000] to indicate the start.[2]
2. When peripherals see that the LFRAME# is asserted, they stop driving the LAD [3:0] lines, even if they are in the middle of a transaction.
3. The host de-asserts LFRAME# and then places information on the LAD [3:0] lines to indicate the cycle type, cycle direction, chip selection, and address. The exact information that is sent by the host depends on the cycle type.[3]
4. The host optionally drives data, and turns the bus around to monitor the peripheral for completion of the cycle.
5. The peripheral indicates completion of the cycle by driving appropriate values on the LAD [3:0] signal lines, and potentially drives data.
6. The peripheral turns the bus around to the host, ending the cycle.

---

[2]Please refer to the "Intel Low Pin Count Interface specification" for start indications for all cycle types.

[3]Refer to the "Intel Low Pin Count Interface specification" for more details.

Table 8.3 Cycles Supported by the LPC Protocol

Cycle Type	Sizes Supported	Comment
Memory read	1 byte	Optional for both LPC hosts and peripherals
Memory write	1 byte	Optional for both LPC hosts and peripherals
I/O read	1 byte	Optional for peripherals
I/O write	1 byte	Optional for peripherals
DMA read	1, 2, 4 bytes	Optional for peripherals
DMA write	1, 2, 4 bytes	Optional for peripherals
Bus master memory read	1, 2, 4 bytes	Optional for both LPC hosts and peripherals, but strongly recommended for hosts
Bus master memory write	1, 2, 4 bytes	Optional for both LPC hosts and peripherals, but strongly recommended for hosts
Bus master I/O read	1, 2, 4 bytes	Optional for both LPC hosts and peripherals
Bus master I/O write	1, 2, 4 bytes	Optional for both LPC hosts and peripherals
Firmware Memory Read	1, 2, 4, 128 bytes	Optional for both LPC hosts and peripherals
Firmware Memory Write	1, 2, 4 bytes	Optional for both LPC hosts and peripherals

From Table 8.3, it can be seen that the Firmware Memory Read cycle can support up to 128 bytes per cycle, whereas all other cycles can only support up to a maximum of 4 bytes per cycle. So, in all cycle types other than the Firmware Memory Read, there is appreciable protocol overhead. Since a TPM module does not require a high bandwidth connection to the SoC, the LPC protocol is adequate.

Power management

In most embedded interfaces, the primary way to reduce power consumption is by stopping the clock signal, and LPC is no different. In low power states, the LCLK signal is stopped, resulting in power savings.

Prior to going into a low power state, the host will assert the LPCPD# signal. The peripherals, on seeing the LPCPD# signal asserted, must make arrangements to drive their LDRQ# signal low or tri-state it. When the LPCPD# signal is de-asserted, peripherals must drive its LDRQ# signal high.

SMART CARD

A smart card is any pocket-sized card containing an embedded integrated circuit. The most familiar form factor for a smart card is the credit-card-sized card. The SIM (Subscriber Identification Module) card, which is found in cellular phones, is also an example of a smart card. Smart cards are sometimes also referred to as Integrated Circuit Cards (ICC) because they have an integrated circuit. The integrated circuit in a smart card can be as simple as a memory; for example, to replace the magnetic stripe in a credit card. Or the smart card can be more complex—with an embedded microcontroller and some memory. Embedding a microcontroller into the smart card allows for extending the functionality of the card. The smart card can then implement enhanced security capabilities such as encryption.

Smart cards can be of two types: contact smart card and contactless smart card. A contact smart card has to be brought into contact with the reader for it to be readable. A contactless smart card can be read by readers without the card being in contact with them. Neither contact nor contactless smart cards have any power supply embedded in them. The contact smart cards derive their power directly from the reader hardware, while the contactless cards derive their power from the reader using radio frequency (RF) induction technology.

The electrical specifications of the smart card, regardless of the form factor, are based on the ISO 7816 specification. The ISO 7816 specification also governs the dimensions and location of the different contacts (Vcc, RST, CLK, and so on) in the smart card.

Security

Encryption is the primary way of ensuring the security of smart card-based transactions. To support this, the smart card needs to have an embedded microcontroller. This controller can then encrypt the data that is sent by the smart card to the reader. The reader is expected to have prior knowledge of the encryption keys that will be used, and will therefore be able to validate and decrypt the data.

Interfaces used

Universal Serial Bus (USB), Personal Computer Memory Card International Association (PCMCIA), and secure digital input output (SDIO) are the most common interfaces used for connecting a smart card reader to an SoC or application processor.

The USB interface is typically used in the case of external smart card readers, where the emphasis is on the reader being hot-pluggable. The USB interface is power hungry, so is not favored for embedded designs. The advantage of USB is its hot-plug ability and widespread availability.

The PCMCIA interface (or one of its derivative interfaces) is typically used to interface smart card readers to a laptop computer. The PCMCIA interface specification is quite old, and the ExpressCard technology is now used in places where PCMCIA was used. In ExpressCard, the host makes use of a PCIe (×1) slot along with a USB 2.0 or USB 3.0 interface. Depending on the usage/application, either or both of these interfaces will be used for performing data transfers.

The SDIO interface is the favored interface for connecting smart card readers in a mobile/embedded host. This is primarily because of its low power consumption. In addition, the SDIO interface also offers data rates up to 100 Mbps, though it is not required for smart card readers.

All three interfaces mentioned have been covered in previous chapters of this book. Since the SDIO interface is most widely used, the specifics of interfacing smart card readers via SDIO are shown in the next section.

SDIO interface

Figure 8.4 shows an SDIO interface to connect a smart card reader to an application processor.

Operation

The SDIO protocol is command-response based. Commands are sent by the host through the command line (CMD line in Figure 8.4). The commands can either be addressed to a particular client or can be broadcast to multiple clients. The client then sends a response through the CMD. The CMD line operates in half-duplex mode. That is, at any given point, the host can send a command or the client can send a response, but both cannot happen together.

The data line(s) are used for transferring data across. SDIO devices can operate in 4-bit or 1-bit mode. As the name implies, in 4-bit mode, the data bus is 4-bits wide, and in 1-bit mode, the data bus is 1-bit wide. It is worth noting that Figure 8.4 shows a 4-bit-wide data line. The data line is bidirectional and also operates in half-duplex mode.

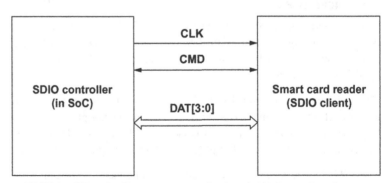

■ **FIGURE 8.4** SDIO interface for connecting smart card reader with 4-bit data transfer mode.

CRC (Cyclic Redundancy Checksum) is the method used for error checking on the SDIO interface. A CRC is generated (and transmitted) for every command, response, and block of data transmitted on the interface. CRCs are helpful in error detection, but are of limited value in correcting the errors. So, in case an error is detected, a retransmission of the data, command, or response is required.

SUMMARY

The recurring theme in this chapter has been the use of encryption to thwart potential attackers from compromising systems where security is paramount.

In the case of NFC and smart card, there are two interfaces that need to be secured:

1. The primary interface. This is the interface through which the client and host communicate with each other; for example, a smart card and a smart card reader. This interface is usually secured through encryption.
2. The interface between the reader and the SoC or application processor. This interface needs to be secured so that the SoC can be sure that the data that is being received is from the correct interface and not from any malicious application executing on the host. This is where interfaces such as Secure I^2C play a part.

In terms of security, TPM (using LPC) has the added advantage that the TPM chip does not initiate any transactions on its own. It only responds to requests from the SoC. So, the TPM chip is not easily compromised.

Power Interfaces

Following up on the general discussion of various interfaces in Chapter 3, this chapter discusses, in detail, the various interfaces used for power delivery, charging, and so on, their applicability to various scenarios, and their capabilities. After this chapter the reader should understand when to use a particular interface and what its advantages are.

VOLTAGE REGULATION

As discussed in Chapter 2, "Understanding Power Consumption Fundamentals," different components of SoC use different power rails. In other words, the whole SoC is divided into various power rails, and different components (of the SoC) operate off these different rails. It is done such that, based on the usage or system operation, only the power rails supporting active components are kept on, and the rest of the rails where no components are active can be switched off. A voltage regulator module (VRM) is a buck converter that provides the SoC components the appropriate supply voltage, converting +5 V or +12 V to a much lower voltage required by the component. A buck converter is a voltage step-down and current step-up converter. The VRMs allow components with different supply voltage to be mounted on the same chip.

The VRMs, when used in non-SoC, processor-only, context, were called *processor power modules*. Based on the design, the VRMs may be soldered onto the motherboard, or they can be installed in an open slot designed especially to accept modular voltage regulators. Recently, the SoC/CPU designers are putting voltage regulation components on the same package (or die) as the SoC/CPU, instead of having a VRM as part of the motherboard. A voltage regulator that is integrated on-package or on-die is usually referred to as a *fully integrated voltage regulator* (*FIVR*) or *integrated voltage regulator*. A FIVR design brings a little bit of simplification to complex voltage regulation.

Operation

The correct supply voltage is communicated by the CPU to the VRM in the beginning via a number called *VID* (voltage identification). Typically, the

VRM initially provides a standard supply voltage to a part of SoC logic whose only aim is to send the VID to the VRM. When the VRM receives the VID identifying the required supply voltage for different components, it starts acting as a voltage regulator, providing the required constant voltage supply to the components.

As an alternative to having a power supply unit generate some fixed voltages, using VID to instruct an on-board power converter of the desired voltage level has several advantages. The switch-mode buck converter adjusts its output according to the VID supplied. This flexibility in generating various desired voltage levels enables delivering a different nominal supply voltage to different components and also different supply voltages at different points in time to the same component, in order to reduce power consumption during idle periods by lowering the supply voltage. For example, a system using 4-bit VID would output 1 of 16 distinct output voltages. These voltages are usually evenly spaced within a given range. Some of the code words may be reserved for special functions such as shutting down the unit, so actually the system with 4-bit VID may have fewer than 16 output voltage levels.

The communication between the VRM and components on SoC is proprietary and varies across various SoC designers/manufacturers.

BATTERY CHARGING

The mobile devices need to run on battery most of the time. So it's imperative for the devices to have an integrated battery. The majority of these devices use rechargeable batteries. The performance and longevity of rechargeable batteries are to a large extent governed by the quality of the charger. The battery and charging module, at first, may appear to be the least significant component because there is no direct functionality provided by them. However, choosing a quality charger is important considering the cost of battery replacement and the frustration poorly performing batteries create. It is not an exaggeration to state that a mobile device is basically useless if the charger doesn't work properly or the battery doesn't last long enough. Fundamentally, batteries have two terminals: one terminal is marked (+), or positive, while the other is marked (−), or negative. Inside the battery case are a cathode, which connects to the positive terminal, and an anode, which connects to the negative terminal. The reaction in cathode and anode form the basis of the current. The reason there are so many different types of batteries is that batteries can use a variety of chemicals to power their reactions. Common battery chemistries include:

- *Zinc-carbon battery:* In zinc-carbon batteries, the anode is zinc, the cathode is manganese dioxide, and the electrolyte is ammonium chloride or zinc chloride. This chemistry is common in many inexpensive AAA, AA, C, and D dry cell batteries.

- *Alkaline battery:* In alkaline batteries, the cathode is composed of a manganese dioxide mixture, while the anode is a zinc powder. This chemistry is also common in AAA, AA, C, and D dry cell batteries.
- *Lithium-ion battery:* A number of substances are used in lithium-ion batteries; however, a common combination is a lithium cobalt oxide as cathode and a carbon anode. These are rechargeable batteries and are typically used in high-performance devices, such as cell phones, digital cameras, and even electric cars.
- *Lead-acid battery:* In lead-acid batteries, the electrodes are usually made of lead dioxide and metallic lead, while the electrolyte is a sulfuric acid solution. This creates rechargeable batteries. This chemistry is typically used in car batteries.

While using a rechargeable battery, we make the chemical reactions run in reverse using a battery charger. Charging up a battery is the exact opposite of discharging it: discharging gives out energy; charging takes energy in and stores it by resetting the battery chemicals to their original state (before being discharged). In theory, the rechargeable batteries can be charged and discharged any number of times; however, in practice, even rechargeable batteries degrade over time and eventually are no longer able to store a charge. This is the point at which we have to recycle the battery.

Fundamentally, all battery chargers have the same mechanism: They feed an electric current through batteries, and the cells inside will hold onto some of the energy passing through them. The cheapest, crudest, and oldest chargers used a constant voltage or constant current and applied that to the batteries until you switched them off. There is no intelligence with the charger; based on how long you keep this plugged in, the batteries may overcharge or remain undercharged. Better chargers use a low, gentle charge for a much longer period of time.

Overcharging is generally worse than undercharging. When overcharged, the batteries heat up and build up pressure inside, which can make them rupture, leak chemicals or gas, and even explode.

So, due to overcharging issues, more sophisticated, timer-based chargers came into existence. These timer-based chargers switched themselves off after a set period. This timer-based mechanism did not completely rule out the possibility of overcharging (or undercharging); however, since the charging time varies for different reasons such as how much charge the battery already had, how old the battery is, and so on, these timer-based chargers definitely did alleviate the repercussions of overcharging and undercharging.

The charging mechanism also did significantly impact the battery lifetime. Since the battery was costly and repercussions of mischarging of batteries were severe, especially in the mobile-handled devices, the next-generation chargers were devised. These next-generation chargers worked intelligently, using a microcontroller-based electronic circuit to sense how much charge is present in the batteries, when the charging is likely to be complete, what the current temperature of the battery is, and then switched off the current or changing to a low trickle charge at the appropriate time to avoid overcharging, overheating, and so on.

USB charging

USB has evolved from a data interface capable of supplying limited power to a primary provider of power with a data interface. Today many devices charge or get their power from USB ports contained in laptops, cars, aircraft, or even wall sockets. USB has become a ubiquitous power socket for many small devices such as cell phones, MP3 players, and other handheld devices.

We already discussed USB in detail in Chapter 5; refer to that chapter for details. In this section, we will recap the key elements that will be used in the USB charging context. The USB standard originally had four signals:

- VBUS: for power supply
- D+ and D− for data differential signaling
- GND

The USB-OTG specification introduced one more signal ID pin for detecting host and device mode of operation.

Since a standard USB downstream port can provide at least 500 mA of current (via VBUS), it was convenient to use it for charging small portable devices (PDs). With the limited current of 500 mA, however, it took quite long to charge the battery. There were extensions to the USB specification to increase the current limit to efficiently charge PDs.

It is important to note that using the standard downstream port (SDP) for charging does not affect or interfere with the regular USB data transfer operation.

There are two standards that define the charging mechanism, current, and voltage while using USB for charging. These two specifications are USB Battery Charging specification (version 1.2) and USB Power Delivery specification (version 1.2). The Battery Charging specification increased the charging current limit from what was provided by the USB 2.0 specification; however, there were still many devices that either required an additional power connection to the wall or exceeded the USB-rated current in order to operate. With international regulations requiring better energy management due to ecological and practical concerns relating to the availability of power, and devices bringing down the power consumption due to the same, USB power delivery attempts to enable the maximum functionality of USB by providing more flexible power delivery along with data over a single cable. The aim of the USB Power Delivery specification is to operate with and build on the existing USB ecosystem and increasing power levels from existing USB standards—for example, battery charging—to enable new higher power use cases such as USB-powered hard disk drives and printers.

Battery charging specification

USB Battery Charging specification was the first attempt to provide reasonable charging capability to a portable device (PD) via USB port. Even though the standard USB downstream port was capable of charging a PD, the charging rate was ridiculously slow due to limited charging current. The Battery Charging specification enables faster charging by means of higher charging current than was permitted by the standard USB specification (USB 2.0 and USB 3.0).

Figure 9.1 shows a simple connection of a PD connected to a standard port or charging port.

Before we get into the details of specification and methodology, we will define a few terms and concepts that are used in the context of charging.

- *Attach*: A downstream device is said to be attached to a USB upstream port when there is a physical cable between the two.

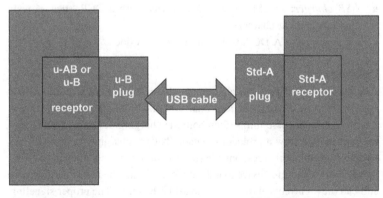

Portable device   **SDP, CDP, or DCP charger**

■ **FIGURE 9.1** A portable device connected to charger.

■ *Connect*: A downstream device is considered connected to a USB upstream port when there is a physical cable between the two, and the device has pulled either D+ or D− high with a 1.5 kΩ resistor.

■ *Dedicated charging port (DCP)*: A USB downstream port that outputs power for battery charging but is not capable of enumerating a downstream device. A DCP must source current (IDCP 0.5-5 A) at an average voltage of VCHG (4.75-5.25 V). A DCP is required to connect the D+ and D− lines with a resistance RDCP_DAT < 200 Ω.

■ *Charging downstream port (CDP)*: A USB downstream port that outputs power for battery charging and complies with the USB 2.0 specification for a USB host or hub downstream port. A CDP must source current (ICDP 1.5-5 A) at an average voltage of VCHG (4.75-5.25 V). If a device is attached but not connected, a CDP must output a voltage of VDM_SRC (0.5-0.7 V) on its D− line if the voltage at D+ is greater than VDAT_REF (0.25 V) and less than VLGC (0.8 V).

■ *SDP*: A USB downstream port that complies with the USB 2.0 specification for a USB host or hub downstream port. An SDP pulls the D+ and D− signals to ground with 15 kΩ resistors and can source up to 500 mA to a configured device. The limit increases to 900 mA if it is a USB 3.0 downstream port.

■ *Accessory charging adapter (ACA)*: An adapter that allows a single USB device to be connected to a charger and another device at the same time.

■ *Dead battery*: A dead battery is defined as a battery with charge so low as to prevent a device from successfully powering up.

■ *PD*: A portable device is a device that is compliant with the USB 2.0 specification and the BC1.2 specification and can draw charging current from USB.

■ *USB charger*: A USB charger is a device with a DCP, such as wall chargers and car chargers.

■ *Charging port*: A DCP, CDP, ACA, or ACA dock.

Overview

Any standard USB port can charge a device if the current required is less than 500 mA (USB 2.0) or less than 900 mA (USB 3.0). If the current required exceeds these limits, then both the charging device and the charging port must follow a protocol to enable battery charging. A downstream battery charging port is responsible for providing the proper handshake signaling to the charging device to indicate that it is attached to a charging port and can draw currents above the standard USB limits. The proper signaling varies depending on the PD. Some PDs follow USB-IF BC1.2 protocols, but there is an installed base of devices that use proprietary handshake protocols, referred to here as legacy modes, for battery charging. It must be noted that in choosing the right charging current the responsibility lies with the PD being charged.

Legacy battery charging

Before the Battery Charging specification came into play, the device/charger manufacturers used their own proprietary mechanism for identifying the charging port or dedicated charger. In other words, legacy devices support some form of battery charging detection intended for use with a dedicated charger. Some of these chargers short $D+$ to $D-$ directly or connect them through a series resistor. For charger detection, some legacy devices assert a voltage on $D+$ by connecting a pull-up and then sensing a voltage on $D-$. If a positive voltage is detected, the device can assume it is plugged into a dedicated charger and not a standard USB port. Other devices pull down one data line while pulling up the other. Once the device detects a charger by the presence of a voltage on $D-$, it can start charging from the VBUS connection at current levels that exceed the USB specification.

Other legacy devices rely on the charger to drive fixed voltages (> 1 V) on the $D+$ and $D-$ data lines; these are referred to as SE1 chargers. If these voltages are sensed by the charging device, the device assumes it is plugged into a dedicated charger and starts charging. A standard USB downstream port would not present these fixed voltages on the $D+$, $D-$ lines.

So, in summary, all these mechanisms were based on assumptions and the tricks were proprietary and these tricks were prone to failure with changes in the USB specification. For example, the devices using 1 V drive voltage on $D+/-$ signal as charger detection may break if the next-generation USB specification version x, uses 1 V signal on $D+/D-$ for regular operations. Due to all these potential and real issues observed, the Battery Charging specification came into play.

USB-IF BC1.2 specification

The USB-IF Battery Charging specification defines current limits and protocols to allow PDs to draw current from USB host ports, hub downstream ports, and dedicated chargers with USB ports in excess of 500 mA (USB 2.0 port) or 900 mA (USB 3.0 port). So the key remains in detecting the charger type and then based on the charger type identified decide the maximum current that can be safely drawn by the device being charged. Attempting to draw 1 A, for example, from a source capable of supplying only 500 mA would not be good. An overloaded USB port will likely shut down, blow a fuse, or trip the switch. Even with resettable protection, it will often not restart until the device is unplugged and reconnected. In ports with less rigorous protection, an overloaded port can cause the entire system to reset.

The key contribution of the USB BC1.2 specification is in standardizing the mechanism for charging port detection. The Battery Charging specification defines a standard mechanism for identifying and differentiating between different types of charging ports, so that an appropriate level of current can be safely drawn by the device being charged. As defined in the USB BC1.2 specification, there are five functional blocks in charger detection logic:

- *VBUS detect*: A PD includes a session valid comparator; VBUS has to be above a threshold voltage before the charger detection is initiated.
- *Data contact detect (DCD)*: This is an optional block used to confirm that the data lines made contact during attachment. A current source on D+ and a pull-down resistor on D− are turned on. If the D + line goes low, this indicates that data lines are attached to a charging port or a standard port, and the logic proceeds to start primary detection. A timeout circuit is required to ensure that primary detection starts after a timeout after attach, in case contact is not detected or the DCD block is not present.
- *Primary detection*: A PD is required to implement primary detection, which is used to distinguish between an SDP and a charging port.
- *Secondary detection*: If a PD is ready for enumeration within a threshold time after VBUS detection it can bypass secondary detection; otherwise it is required to implement secondary detection. Secondary detection is used to distinguish between a DCP and a CDP port.
- *ACA detection*: ACA detection support for a PD is optional, and only PD devices with a USB micro-AB connector can support ACA detection, because detection is done by measuring the resistance of the ID pin.

Once the charging port type is identified, the available current for the various ports is as follows:

1. DCP: 500-1.5 A
2. CDP (host or hub): to 900 mA (580 mA during chirp) for Hi-Speed; to 1.5 A for low and fast speed
3. Low-power SDP (host or hub): 100 mA
4. High-power SDP (host or hub): 500 mA

The available current can be used by the battery or the system, or it can be split between them.

USB Power Delivery specification

The USB Power Delivery specification attempts to enhance the maximum allowed current that can be used by the device being charged. The specification describes the architecture, protocols, power supply behavior, connectors, and cabling necessary for managing power delivery over USB at up to 100 W. The USB Power Delivery specification defines a power delivery system covering all elements of a USB system, including hosts, devices, hubs, and chargers. The specification is intended to be fully compatible and to extend the existing USB infrastructure. USB power delivery is designed to operate independently of the existing USB-defined mechanisms currently used to negotiate power, which are: in band requests for high-power interfaces in USB 2.0 and USB 3.0, and mechanisms for supplying higher power as defined in Battery Charging specification 1.2. Initial operating conditions remain the default USB states as defined in USB 2.0, USB 3.0, or BC1.2, that is:

- The downstream port sources vSafe5V over VBUS.
- The upstream port consumes power from VBUS.

The USB Power Delivery specification is intended as an extension to the existing USB 2.0, USB 3.0, and BC1.2 specifications. To give a perspective, Table 9.1 compares the charging current and voltage between the USB 2.0 and USB-PD.

Overview

In USB power delivery, pairs of directly attached ports negotiate voltage, current, and/or direction of power flow over the power conductor (VBUS). The mechanisms used operate independently of other USB methods used to negotiate power. Any contract negotiated using this specification supersedes any and all previous power contracts established whether from standard

Table 9.1 Comparison of Charging Current and Voltage Between USB 2.0 and UPD

SPEC	USB 2.0	UPD
Voltage (V)	5	20
Current (A)	0.5	5
Power (W)	2.5	100

[USB2.0], [USB3.0], or [BC1.2] mechanisms. Once a contract is negotiated, the port pair remains in power delivery mode until the port pair is detached or goes through Hard Reset.

When PD devices are attached to each other, the downstream and upstream ports initially default to standard USB operation. The downstream port supplies 5 V and the upstream port draws current in accordance with the rules defined by [USB2.0], [USB3.0], or [BC1.2] specification. After power delivery negotiation has taken place, power can be supplied at higher, or lower, voltages and higher currents than defined in these specifications. It is also possible to swap the power supply roles such that the downstream port receives power and the upstream port supplies power.

A USB power delivery port supplying power is known as a *source* and a port consuming power is known as a *sink*. There is only one source port and one sink port in each PD connection between port partners. In USB power delivery protocols a USB downstream port is initially a source and a USB upstream port is initially a sink, although USB power delivery also enables these roles to be swapped.

Each USB power delivery capable device is assumed to be made up of at least one port. Providers are assumed to have a source and consumers a sink. Each device contains one or more of the following components:

- Upstream ports that may contain
  - sink power (a consumer),
  - optionally source power (a consumer/provider), and
  - optionally communicate via USB.
- Downstream ports that may contain
  - source power (a provider),
  - optionally sink power (a provider/consumer), and
  - optionally communicate via USB.
- A source that may be
  - an external power source such as AC,
  - power storage (such as a battery), and
  - derived from another port (such as a bus-powered hub).
- A power sink that may be
  - power storage (such as a battery),
  - used to power internal functions, and
  - used to power devices attached to other devices (such as a bus-powered hub).

Architectural overview

This section outlines the high-level logical architecture of USB power delivery specification. In practice various implementation options are possible based on many different possible types of power delivery based on the topology. For example, PD devices may have many different configurations, such as USB or non-USB communication, single versus multiple ports, dedicated power supplies versus supplies shared on multiple ports, or hardware- versus software-based implementations. The architecture in this section therefore is provided only for reference in order to indicate the high-level logic model used by the USB-PD specification. This architecture is used to identify the key concepts and also to indicate logical blocks and possible links between them.

The USB power delivery architecture in each USB power delivery capable device is made up of a number of major components. The communications stack consists of:

- *Device policy manager*: The device policy manager is a mandatory component, and it exists in all devices and manages USB power delivery resources within the device across one or more ports based on the device's local policy.
- *Policy engine*: The policy engine exists in each USB power delivery port and is responsible for implementation of the local policy for that port.
- *Protocol layer*: The protocol enables messages to be exchanged between a source port and a sink port.
- *Physical layer*: The physical layer handles the actual transmission on the wire.

As expected, the physical layer sits at the bottom and the device policy manager at the top; the policy engine and protocol layer sit in between. The entire communication stack sits over the VBUS, which is the only real connection between the provider and the consumer.

Additionally, USB power delivery devices that can operate as USB devices may communicate over USB. An optional system policy manager that resides in the USB host communicates with the PD device over USB, via the root port, and potentially over a tree of USB hubs. The device policy manager interacts with the USB interface in each device in order to provide and update PD-related information in the USB domain. It must also be noted that a PD device is not required to have a USB device interface.

Figure 9.2 shows an example USB connection arrangement and respective PD topology. It must be noted that some pieces can play the role of provider and consumer both while the others may have an exclusive role of either provider or consumer.

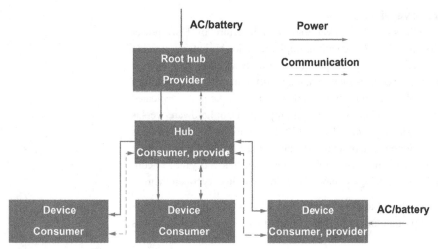

■ **FIGURE 9.2** An example USB connectivity arrangement and respective PD topology.

SUMMARY

So, to summarize the discussion in this chapter: The voltage regulators (VRs) are used for supplying voltage to various components in the SoC without requiring separate supplies for different components. Voltage regulators (VRs) also enable effective power management. The voltage regulation-related communication happens using VID and is proprietary.

Batteries are essential components of mobile devices, and effectively and properly charging the batteries is important since overcharging or overheating can be fatal to the system and/or injurious to the user. The battery chargers moved from no intelligence to becoming intelligent. USB, being the most prevalent interface for data communication, has been extended with the Battery Charging and Power Delivery specification to deliver charging power effectively.

Chapter 10

Sensor Interfaces

The interfaces through which sensors are commonly connected to an system on chip (SoC) are explained in this chapter. All of these interfaces have already been explained in detail in preceding chapters. This chapter provides a brief overview of each interface to refamiliarize the reader and proceeds to lay out the reasons for choosing one interface over another in connecting a sensor. In this chapter, the term *sensor* includes touch panels, motion sensors (accelerometers, gyroscopes), and environmental sensors (ambient light sensors, barometer, and so on).

INTER-INTEGRATED CIRCUIT

The I^2C (Inter-Integrated Circuit) bus is used for connecting low-speed peripherals to an SoC. It makes use of two lines for data transfer—a data line (serial data [SDA]) and a clock line (serial clock [SCL]). Data is transferred serially over the I^2C bus. Every I^2C bus configuration can have one or multiple masters and one or multiple slaves.[1] Each slave has an I^2C address that is used by the master to communicate with that particular slave.

As can be seen from Figure 10.1, multiple I^2C slaves can be connected to the same master. Transactions are always between the master and one of the slaves.[2] Transactions are unidirectional—read transactions, where the slave transmits data to the master, and write transactions, where the master transmits data to the slave.

Operation

All transactions are initiated by the master. To begin a transaction, the master generates a START condition on the bus and then transmits the address of the slave that it wishes to communicate with, along with an indication of the type of transaction (read/write). The slave, if it is present, sends an

---

[1]Multiple master I^2C configurations are rarely used. A typical I^2C bus configuration will be one master and *n* slaves (where $n > 1$).

[2]Strictly speaking, this is not true. There is a provision for General call where the master can broadcast a message to all the slaves on the bus that support General call.

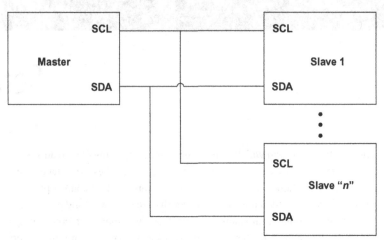

■ **FIGURE 10.1** An example I²C bus configuration.

acknowledgment to the master. This begins the transfer. After receiving each byte, the receiver acknowledges it. The sender waits for this acknowledgment before transmitting the next byte. The receiver does not acknowledge the last byte in a transaction; this indicates to the sender that no more data is to be sent.

Table 10.1 shows the most common modes at which an I²C bus operates and the corresponding clock speeds.

Since there is a bit of overhead introduced by the I²C protocol with its addressing and the acknowledgments following each transmitted byte, the throughput is a bit lower than the clock rate. In case the transaction lengths are short, the overhead will be quite large, since the address byte needs to be transmitted for each transaction. Larger transaction lengths have the lowest overhead and by extension give the best throughput.

Table 10.1 I²C Modes of Operation

I²C Mode	Bus Speed
Standard Mode	100 kHz
Fast Mode	400 kHz
Fast Mode +	1 MHz
High Speed Mode	3.4 MHz
Ultra-Fast Mode	5 MHz

Here is a brief calculation indicating the throughput achievable with a bus speed of 5 MHz (Ultra-Fast Mode). Assume all transactions are x bytes long, there is no clock stretching, and the addresses are 7-bit addresses.

$$\text{Bits to transmit} = 8 \times x$$
$$\text{Address byte} = 8 \text{ bits}$$
$$\text{Acknowledgement bits} = 1 + (x - 1)$$

(The address byte is acknowledged, but the last byte to be transmitted is not acknowledged.)

$$\text{Total bits (equal to clock cycles)} = 8x + 8 + x$$
$$= 9x + 8$$

The "total bits" figure also represents the total number of clock cycles required for the transfer. Note that the calculation assumes that there are no delays between transmitting 2 bytes for simplicity. In practice, there will be delays introduced by both hardware and software.

Now, let us define effective utilization as the ratio between the clock pulses that are used for transmitting actual data and the total clock pulses.

For $x = 1$ (remember x is the total number of bytes per transfer):
Effective utilization $= (8 \times 1)/((9 \times 1) + 8) = 8/17 = 47.06\%$
For $x = 8$:
Effective utilization $= (8 \times 8)/((9 \times 8) + 8) = 64/80 = 80\%$

At 80% effective utilization, an I^2C bus operating at 5 MHz (Ultra-Fast Mode) will yield a throughput of 4 Mbps. This is quite low compared to other serial protocols, such as the Serial Peripheral Interface (SPI), which we will see next.

Usage for sensors

The key advantage of the I^2C protocol is that it requires only two pins on the master, irrespective of the number of slaves that are connected to it. The number of slaves connected to a master is limited only by the number of unique I^2C addresses that can be used. For 7-bit addressing, this works out to a total of 112[3] slaves that can be connected to a single master. This number will be even higher if 10-bit addressing is used. Needless to say, this limit is never approached, and typical I^2C bus configurations will have anywhere between 1 and 10 slaves.

---

[3]Seven address bits results in $2^7 = 128$ possible combinations, out of which 16 addresses are reserved. This results in $128 - 16 = 112$ addresses being available for usage by slave devices.

Another advantage of the I^{2}C protocol is that it is a simple protocol and relatively easy to implement in software. Though an I^{2}C controller is implemented in most entry-level microcontrollers, there may be some cases where an I^{2}C controller is not available. One reason could be that there may be too many I^{2}C slaves and not enough I^{2}C interfaces on the microcontroller.[4] In such cases, two spare general purpose input/output (GPIO) lines can be designated as the I^{2}C clock and data lines, and the I^{2}C protocol can be implemented wholly in software. This is known as bit-banging.

The I^{2}C protocol is also widely supported in software, at least in the context of sensors. Both the Windows 8 and Linux operating systems have support for the HID-over-I^{2}C protocol. Here, HID stands for Human Interface Device and sensors fall under this category. In the most typical case, a number of sensors and a microcontroller are combined together to form a standalone entity (we will call it the sensor hub going forward). The sensor hub's microcontroller implements support for the HID-over-I^{2}C protocol and is connected to the application processor via an I^{2}C interface. With native support for the HID-over-I^{2}C protocol in an operating system, the sensor works as is without the need for a driver.

See Figure 10.2 for a sample representation of a sensor hub-based configuration. One thing to note is that the connection between the microcontroller and the sensors within the sensor hub can be through any interface (I^{2}C/SPI

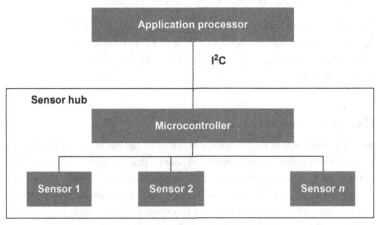

■ **FIGURE 10.2** A sensor hub-based system configuration.

---

[4]One other example: There may be one critically important I^{2}C slave and some other not-so-critical I^{2}C slaves on a system that has a microcontroller with a single I^{2}C interface. In this case, the critically important slave is put on the I^{2}C interface, and the other slaves are put on the bit-banged I^{2}C interface.

is typically used here). This does not affect the use of HID-over-I^2C for communication between the sensor hub microcontroller and the application processor.

One drawback of the I^2C protocol is its low speed. The I^2C interface is unsuitable for sensors that produce lots of data (such as a fingerprint sensor). Most of the commonly used sensors—accelerometers, gyroscopes, magnetometer, and the like—comfortably fit within the bandwidth afforded by the I^2C protocol.

I^2C protocol dictates that the clock and data lines should be pulled up high when not in use. This results in tradeoff between the bus speed and the power consumption. Lower resistance pull-ups will increase the bus speed, but also increase power consumption. Higher resistance pull-ups will lower the bus speed, but decrease power consumption.

Sensors most suited to connecting via I^2C: Touch panels, accelerometers, gyroscopes, magnetometers, environmental sensors (ambient light sensors, pressure sensors), and proximity sensor.

SERIAL PERIPHERAL INTERFACE (SPI)

SPI is a full-duplex interface that uses four lines for communication. For this reason, SPI is sometimes called a four-wire serial bus. The four wires are

1. SS: Slave Select
2. SCLK: Serial Clock (driven by master)
3. MOSI: Master Out Slave In (data going from master to slave)
4. MISO: Master In Slave Out (data going from slave to master)

A typical SPI bus configuration is shown in Figure 10.3.

The SS line is used by the master to select the slave that it wishes to communicate with. One SS line is required per slave. So, if n slaves are connected to a master in an SPI bus configuration, n SS pins are required on the master. The SCLK, MOSI, and MISO lines are shared between all slaves. For a bus configuration where n slaves are connected to a master, $(3+n)$ pins will be required on the master.

Operation

Each transaction begins with the master selecting a slave using the SS line. The master then proceeds to operate the SCLK at a frequency less than or equal to the maximum frequency supported by the slave (typical speeds are in the MHz range). In each clock cycle, the master sends 1 bit to the slave,

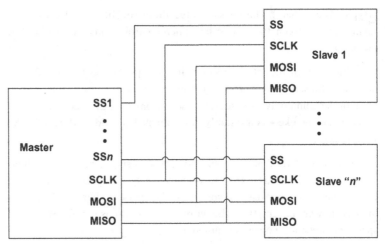

■ **FIGURE 10.3** An example SPI bus configuration.

and the slave sends 1 bit to the master. In case either the master or the slave has no data to send, they can send 0 s.

There is no possibility for flow control in this protocol. The master controls the clock, so it can potentially use that to implement some form of flow control, but it is rarely done. The slave has no means of implementing/enforcing flow control and is expected to process all the data that it receives.

There is no standard set of speeds defined for the SPI protocol. Typical bus speeds are in the 50 MHz range. Since there is no overhead added by the protocol such as addressing and flow control, the throughput that can be achieved using SPI mirrors the clock frequency. For a 50 MHz SPI line, the throughput is 50 Mbps. One more thing to remember is that the communication is full duplex. So, the effective throughput in the case mentioned can be as high as 100 Mbps (if both the master and slave are able to send meaningful data to each other).

Usage for sensors

The key advantage of the SPI protocol is that it supports very high speeds, and by extension, throughput. Typical SPI bus speeds are in the 50 MHz range and with the absence of any protocol overhead as in the case of I^2C, this translates exactly into bit rate. This makes the SPI bus most suitable for sensors that produce a lot of data, such as, for example, fingerprint sensors.

Full-duplex communication is enabled in the SPI protocol by default. If there is a situation where both the master and slave need to transmit equal volumes of data to each other, this feature can be employed.

The SPI protocol is comparable in complexity to the I^2C protocol—they are both quite simple. In SPI, sampling a data line and changing the data values must both be synchronized to the rising and falling edges of the clock line. Bit-banging the SPI protocol using GPIO lines is also common, as is the case with I^2C.

With the absence of pull-up resistors that are present in the case of I^2C, the power consumption of an SPI bus interface will be marginally lower than that of an I^2C interface.

Since there is no flow control mechanism embedded into the protocol for SPI, the software will have to process the received data faster than it arrives. This problem is mitigated by the presence of first in first out (FIFO) buffers on the receiving end. There is no error-detection and correction provision embedded into the protocol for SPI. If required, error detection/correction has to be implemented in software.

There is no in-band addressing in SPI (unlike I^2C). Each SPI slave that is connected to a master will need a dedicated SS pin on the master. This means the number of pins required on the master will increase as the number of connected slaves increases. It is typical for a SPI master to control two to three slaves and not more.

There is also no formal standard for the SPI protocol, so there is no benchmark for validating protocol conformance of a given product.

Sensors most suited to connecting via *SPI*: Fingerprint sensors and touch panels.

In addition to this, certain display controllers use SPI for accepting configuration parameters. Some flash memory also uses SPI.

UNIVERSAL ASYNCHRONOUS RECEIVER TRANSMITTER (UART)

UART (Universal Asynchronous Receiver Transmitter) is one of the oldest protocols used for serial communication. In this protocol, data is transmitted and received via two lines: one each for transmit and receive. A UART link can operate in the following modes:

- *Simplex*: Data transmission is always in one direction. The receiving device has no way of sending any data.

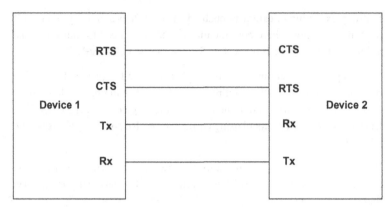

■ **FIGURE 10.4** An example UART bus configuration.

- *Half duplex*: Data can flow in both directions, but only in one direction at a time.
- *Full duplex*: Data flows in both directions simultaneously.

In a UART bus configuration, two devices are connected to each other. There is no provision for more than two devices to be connected on the same bus. Please see Figure 10.4 for a sample UART bus configuration.

There is a key difference between the serial protocols discussed earlier in this chapter (I^2C and SPI) and UART. UART is an asynchronous protocol while I^2C and SPI are synchronous protocols. In a synchronous protocol, the clock signal is transmitted along with the data, so it is easy for the receiving end to decode the data received. In an asynchronous protocol, the clock signal is not transmitted along with the data.

To understand the impact of this difference, take a look at the waveform shown in Figure 10.5. Assume that the data line is to be sampled when the clock line is high.

In the example in Figure 10.5, a high signal represents logic 1 and a low signal represents logic 0. When the clock signal is transmitted along with the data signal, the receiving end is able to decode that the data transmitted was [1 0 00].

Now, if the clock signal is not transmitted, the receiver will find it difficult to decode the transmitted data—was it [1 0 0], [1 0 00], [1 1 0 0 0], or [1 0 00 0]? This is because without the aid of a clock signal, the receiver cannot know how many 1 s and 0 s were transmitted.

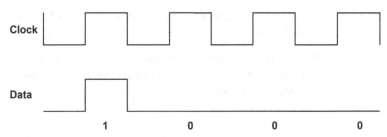

Clock

Data

1   0   0   0

■ **FIGURE 10.5** Example waveform of transmission with separate data and clock.

The UART protocol has the following features to get around this problem:

- Use of a start bit to signal the start of transmission.
- The receiver knowing the bit rate at which the transmitter operates.

Operation

This section assumes that the Ready To Send (RTS) and Clear To Send (CTS) lines in the transmitting and receiving devices are cross-connected (as shown in the example of Figure 10.4). When a device is ready to receive data, it sets the RTS line. Before transmitting data, a device must check to determine whether the CTS line is set (the CTS line being set corresponds to the receiver asserting the RTS line), and if yes, proceed to transmit.

Each character is coded as shown in Figure 10.6 in the UART protocol.

The transmitter starts things off with a start bit. The data lines are pulled high when the bus is inactive, and for the start bit, the data line is pulled low for 1 bit duration. After sending the start bit, the transmitter proceeds to transmit the data bits. The number of data bits is configurable and is between 5 and 8. After the data bits, an optional parity bit for error detection is appended. Following the parity bit, the stop bit(s) are transmitted. The number of stop bits is configurable. It can either be 1 or 2.

The receiver uses the start bit for synchronizing its clock to that of the transmitter. As already mentioned, the receiver knows the bit rate at which the transmitter operates. It has an internal clock that operates at a multiple (typically $8\times$ or $16\times$) of the bit rate. At each clock pulse, the receiver samples the data line. If a transition in the data line lasts for less than one-half of the

Bits	0	1-8	9	10-11
Contents	Start bit	Data bits	Parity bit (optional)	Stop bits

■ **FIGURE 10.6** UART character framing.

bit duration, it is considered a spurious pulse. All other transitions of the data line are considered valid.

While simplistic UARTs use only the start bit for clock synchronization, more complex UARTs use all line transitions that are not considered spurious pulses for clock synchronization.

Usage for sensors

The UART protocol typically allows for only two devices to be connected to each other.[5] There is no easy way of connecting multiple UART devices to a single interface. Just for information, one of the ways to accomplish this is by connecting multiple UART peripherals to a single interface through a switch and then using some form of time-multiplexing to switch between the different peripherals. This is rarely done in practice.

UART interfaces have a maximum data rate of around 5 Mbps. There is also some protocol overhead in the form of start, stop, and parity bits. The data rate of a UART interface is similar to that of an I^2C interface.

With UART being an asynchronous interface, it does not have the clock-skew-related problems that synchronous interfaces like I^2C and SPI have. In synchronous interfaces the clock is transmitted along with the data and is required for decoding the received data. If the clock and data signals go appreciably out of phase, this will result in the received data being decoded wrongly. This is one of the main reasons why I^2C and SPI work best for short links (in the order of a few inches), because longer links are more likely to introduce phase differences between the clock and data lines. Synchronous interfaces like UART do not have this problem because there is no clock line and by extension, no clock skew. UART interfaces can operate over much longer cable lengths, usually in the order of a few meters.

To compensate for the clock signal not being transmitted, the transmitting and receiving ends of the UART interface must agree on the protocol parameters (start, stop, parity bits) and the bit rate before the transfer is initiated. In addition, the transmitter and receiver should have stable oscillators with minimal jitter. This is required for proper clock synchronization.

The UART protocol has support for flow control (RTS/CTS) and error detection (parity bits) built into it. In case error correction is required, it must

---

[5]This is true for RS-232, which is the most commonly used signaling protocol for UART. There are other signaling interfaces, such as RS-422 and RS-485, which allow for single/multiple master and multiple slave configurations.

be implemented in software. Cyclic redundancy check (CRC) checksums are the preferred way of doing it, but are very resource intensive.

Since the most typical UART bus configuration consists of two devices connected to each other, it is not generally used for connecting sensors. I^2C and SPI are preferred over UART since they allow multiple sensors to be connected to the same master. UART interfaces are widely used for connecting Bluetooth and GPS chips to an application processor.

UNIVERSAL SERIAL BUS (USB)

The Universal Serial Bus (USB) protocol is used for connecting a wide variety of devices to a processor/SoC. Some example devices that can be connected via USB are mouse, optical drives, flash memory, touch screens, and webcams. The USB protocol is less prone to interference than the other protocols described earlier in this chapter because it uses differential signaling.

DIFFERENTIAL SIGNALING

When a signal is transmitted through a single wire, surrounding electrical interference acts on it and can potentially modify the signal levels, causing the signal to be detected incorrectly (a 1 getting detected as 0 and vice versa).

In differential signaling, each signal is transmitted using a differential pair—the signal carried by one wire is the same level as the one carried by the other wire, but in opposite polarity. The signal at the receiving end is interpreted as the difference between the two lines that make up the differential pair.

If interference acts on the differential pair, it modifies both the lines similarly, but does not affect the difference between the lines. This makes differential signaling immune to electrical interference.

The architecture of USB topology is asymmetrical. It consists of a host, one or many downstream USB ports, and peripheral devices connected to ports in a tiered-star topology. Additional USB hubs may be included in the tiers. Inclusion of hubs allows for branching into a tree structure. The maximum number of tier levels are five. A USB host may implement multiple host controllers and each host controller may provide one or more USB ports. The maximum number of devices, including hub devices, which can be connected to a single host controller is 127. There is one hub that is built into the USB host controller that is called the root hub.

A physical USB device may consist of several logical subdevices that are referred to as device functions. A single device may provide several functions; for example, a webcam (video device function) with a built-in microphone (audio device function). This kind of device is called a composite device. An alternative to this is a compound device, in which the host assigns each logical device a distinctive address and all logical devices connect to a built-in hub that connects to the physical USB cable.

USB device communication is based on pipes (logical channels). A *pipe* is a connection from the host controller to a logical entity, found on a device, and named an *endpoint*. Because pipes correspond 1-to-1 to endpoints, the terms are sometimes used interchangeably. A USB device could have up to 32 endpoints (16 in, 16 out), though it is rare to have so many. An endpoint is defined and numbered by the device during initialization (the period after physical connection called *enumeration*) and so is relatively permanent, whereas a pipe may be opened and closed.

Operation

The USB protocol has been explained in detail in Chapter 5. This section just serves to refresh the most significant points.

USB communication is always between two kinds of devices: host and device. The USB protocol is packet based, and all transactions begin with a token packet. The token packet is generated by the host to describe what is to follow and whether the data transaction will be a read or write and what the device's address and designated endpoint is. Data packets are used to transmit the actual data. Status packets are used for acknowledging transactions, and for error correction.

The packets that comprise a transaction vary depending on the endpoint type. The following are the types of endpoints:

- *Bulk*: Bulk transactions are characterized by the ability to guarantee error-free delivery of data between the host and a function by means of error detection and retry.
- *Control*: Control transfers identify, configure, and control devices. They enable the host to read information about a device, set the device address, establish configuration, and issue certain commands.
- *Interrupt*: Interrupt transfers are typically nonperiodic, small device "initiated" communication requiring bounded latency. An interrupt request is queued by the device until the host polls the USB device asking for data.
- *Isochronous*: Isochronous transfers occur continuously and periodically. They typically contain time-sensitive information, such as an audio or video stream.

Usage for sensors

The power consumption of USB is an order of magnitude higher than the other interfaces discussed in this chapter—both in Active and Idle/Suspended modes. For this reason, the USB interface is not favored for connecting sensors in SoC-based designs.

Since the USB-based devices (for that matter, all devices) on an SoC platform will spend a majority of time in the idle state, the increased power draw during idle state is the bigger concern, rather than the power consumption during active state. Typically, USB devices enter a state called *selective suspend* when the device is idle. This is the deepest sleep state defined by the USB specification. In the selective suspend state, the USB device goes into suspend state. The bus is still powered. This is the reason for the increased power consumption during suspend mentioned earlier.

To further decrease the power consumption of USB devices during idle/suspend state, certain USB devices add support for a state known as D3cold. In that state, the system cuts power to the device, the device drops off the bus, and the device does not consume any power. However, the caveat is that the device cannot wake the system from suspend state, since it is powered off. In case a wake-capable device needs to implement D3cold, the platform designer must implement an out-of-band wake mechanism for that device, for example, using GPIO lines.

In terms of performance, the USB protocol supports speeds up to 5 Gb/s (for USB 3.0) and 10 Gb/s (for USB 3.1). This makes USB one of the best protocols in terms of performance.

Similar to the HID-Over-I^{2}C specification, there is also an HID-Over-USB specification.[6] The HID-Over-USB specification also has native support in the Windows 8 and Linux operating systems. It follows that if a hardware device is designed to support the HID-Over-USB specification, it will work without the need for a separate driver.

USB-based sensor evaluation boards are very popular among developers. Due to the widespread software support for USB and the availability of a USB port on most personal computers, it is easy for a developer to quickly prototype a sensor design on a USB-based development board. However, due to the limitations mentioned, production designs make use of other interfaces for connecting sensors.

---

[6]In fact, the HID-Over-USB specification predates the HID-Over-I^{2}C specification.

Sensors typically connected via *USB*: Touch panels, digitizers (pen input), though I²C is the preferred interface for connecting both of them.

SUMMARY

Table 10.2 provides a side-by-side comparison of the different interfaces explained in this chapter.

In SoC-based designs, having a low pin count is desirable. The I²C and SPI interfaces, with their ability to support multiple slave devices in a single controller, are quite useful in this regard. In terms of power consumption, they both consume low power, with SPI having even lower power consumption than I²C. Not surprisingly, these two interfaces are the most widely chosen ones for connecting sensors.

The UART interface is useful in situations where the link length needs to be slightly longer than usual, for whatever reason. With its asynchronous signaling, UART links are less prone to clock skew, which results in larger link lengths. In terms of speed and power consumption, UART is comparable to I²C and SPI. However, since only one UART slave can be connected to one controller, UART uses up more pins on the SoC than I²C or SPI.

The USB interface is favored for performance, but accomplishes this at the cost of power. For this reason, it is not widely used in SoC-based designs for connecting sensors.

Table 10.2 Comparison of Typical Interfaces Used for Interfacing Sensors

	I²C	SPI	UART	USB
Speed	Low	Medium	Low	High
Power consumption	Low	Very low	Low	High
Software support	High	Moderate	Moderate	High
Availability of standards	Yes	No	Yes	Yes
Link length	Short	Short	Long	Long

Input Device Interfaces

This chapter describes the interfaces that are commonly used for input devices such as mouse, keyboard, and remote control. The details provided for each interface are expected to guide the user on selecting the appropriate interface based on the user's requirements. The touch panel, which can also be considered as an input device, has already been discussed in Chapter 10.

KEYBOARD

The keyboard is the primary interface for inputting alphanumeric characters into computers. The interface used for a keyboard need not have high bandwidth capability due to the low volume of data generated by a keyboard. However, the transfer latency should be minimal so that users get an instantaneous response when they press a key.

The typical interfaces used to connect a keyboard to a processor are Bluetooth and Universal Serial Bus (USB), with USB being more widely used. The DIN5 (now obsolete) and PS/2 (considered legacy and not used widely) interfaces were used in the past to interface keyboards—they are covered briefly for the sake of completeness.

DIN5

DIN5 (or DIN 41524) is a set of standards defined by the Deutsches Institut für Normung (DIN), which is the German national standards organization.

The DIN5 interface transmits keystrokes serially over the data line. Apart from this, there are Vcc, clock, and ground lines. The DIN5 interface was replaced by the PS/2 interface in the early 1990s.

PS/2

The PS/2 keyboard connector replaced the DIN5 connector for connecting keyboards. The name comes from the IBM Personal System/2 series of personal computers, with which it was introduced in 1987. The PS/2 protocol is electrically similar to the DIN5 protocol discussed earlier.

Protocol overview

The PS/2 interface is idle when both the data and the clock lines are pulled high. This is the only state when the device (keyboard, mouse, and so on) can begin transmitting data. The host[1] has ultimate control over the bus and may inhibit communication at any time by pulling the clock line low. The device always generates the clock signal. If the host wants to send data, it first inhibits communication from the device by pulling clock low. After that the host pulls data low and releases clock. This is an indication to the device to start generating clock pulse. The state is referred to as "Request-to-Send" state. Table 11.1 shows various states the PS/2 interface bus can be in and the state of data and clock lines during those states.

All data is transmitted 1 byte at a time and each byte is sent in a frame consisting of 11-12 bits. These bits are

- 1 start bit. This is always 0.
- 8 data bits, least significant bit first.
- 1 parity bit (odd parity).
- 1 stop bit. This is always 1.
- 1 acknowledge bit (used only in host-to-device communication).

Data sent from the device to the host is written on the data line when the clock is high and read on the falling edge of the clock signal. Data sent from the host to the device is written when the clock is low and read on the rising edge. The clock frequency must be in the range 10-16.7 kHz.

PS/2 drawbacks

The PS/2 interface has the following drawbacks:

1. PS/2 interfaces are designed to directly connect the digital I/O lines of the microcontroller on the device with digital I/O lines on the host. This

Table 11.1 Bus States for the PS/2 Interface

Data Line	Clock Line	Bus State
High	High	Idle state
High	Low	Communication inhibited
Low	High	Request to send (from host)

---

[1]The term *host* has been used in this chapter interchangeably with SoC or application processor. It simply refers to the main processor in a system.

theoretically means that the PS/2 interface is not designed to be hot swappable. Though hot-swapping might work if the device being swapped out is very similar to the one being swapped in, it is not recommended.

2. PS/2 connectors, with their slim pins, are not designed to be plugged in and out multiple times. Doing so is likely to damage the connector.

3. The PS/2 interface can be used to connect both a mouse and a keyboard to the same microcontroller on a host. While this brings the cost down, on rare occasions a misbehaving device has been known to bring the entire interface down. For example, a faulty mouse may cause the keyboard to stop functioning.

The USB interface, which solves most of the drawbacks of the PS/2 interface, is the currently preferred method of connecting a keyboard to a host.

Universal Serial Bus (USB)

USB protocol is used for connecting a wide variety of devices to a processor or SoC. Most modern keyboards are connected via USB. Some examples of other devices that can be connected via USB are mouse, optical drives, flash memory, and webcams.

The USB protocol has been discussed in detail in the preceding chapters, so only the details that are required for understanding the keyboard interface are presented here. Consult Chapter 5 for details on the USB protocol.

Operation

USB keyboards are classified as human interface devices (HIDs). They are expected to follow the HID over USB protocol. Most operating systems have built-in support for the protocol and USB keyboards that are HID compliant (which is the vast majority, if not all of them) will work without having to install any device drivers on the host machine.

In USB, there are one host and multiple devices connected via a tiered star topology. There are logical entities on USB devices that are named endpoints. The host communicates via pipes, which are logical channels, to the endpoints on the devices.

In USB terminology, the direction of an endpoint is based on the host. Thus, IN always refers to transfers to the host from a device and OUT always refers to transfers from the host to a device. USB devices can also support bidirectional transfers of control data. Figure 11.1 shows a USB-based keyboard interface.

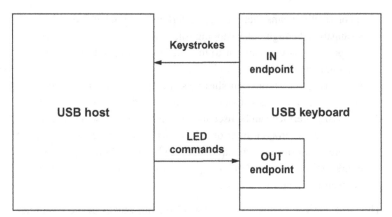

■ **FIGURE 11.1** USB device (keyboard) connected to host.

Here, the IN endpoint is used to send keystrokes to the host, while the OUT endpoint is used by the host to let the keyboard know which of its light-emitting diodes (LEDs) must be turned on.

The USB interface overcomes all the disadvantages of the PS/2 interface:

1. USB devices are designed to be hot swappable.
2. The USB connector is quite robust, and there are no restrictions on the number of times it can be plugged in and out.
3. A USB device may stop working, though it is quite rare. Since USB devices are hot pluggable, removing it and plugging it back in usually causes the driver to get reloaded and the device to start working.
4. In any case the kind of problems seen in PS/2 where a faulty device can bring down the entire interface will not happen in USB.

Bluetooth

Bluetooth, which was originally thought of as a substitute for the RS-232 serial interface, is a wireless communication standard through which devices can communicate over short distances. Bluetooth data rates can go up to a theoretical maximum of 3 Mbps, but this speed is never reached in practice, nor is it required for the kind of usage models supported by Bluetooth. The range offered by Bluetooth depends directly on the transmission power. Bluetooth devices are divided into three classes based on the transmission power, as shown in Table 11.2.

A majority of Bluetooth devices, including keyboards and mice, belong to device class 2.

Table 11.2 Bluetooth Device Classes

Device Power Class	Maximum Output Power (mW)	Typical Range (m)
Class 1	100	100
Class 2	2.5	10
Class 3	1	1

Interference

Bluetooth technology operates in the unlicensed industrial, scientific, and medical (ISM) band at 2.4-2.485 GHz. The 2.4 GHz ISM band is available and unlicensed in most countries. The Bluetooth technology uses spread spectrum, frequency hopping, full-duplex signaling at a nominal rate of 1600 hops/s.

The Wi-Fi protocol also operates in the 2.4 GHz[2] frequency range. In addition, many cordless telephones also operate in the same frequency range. Surprisingly, the USB 3.0 computer cable standard has been proven to generate significant amounts of electro magnetic interference (EMI) that can interfere with any Bluetooth devices a user has connected to the same computer.

To minimize interference, Bluetooth makes use of adaptive frequency hopping technology. In this, the frequency band available for Bluetooth is divided into 79 channels, each 1 MHz wide. The Bluetooth devices then listen in to map the frequencies at which other devices in the vicinity operate, and choose the frequencies where there is no interference for their operation.

Operation

Bluetooth is a packet-based protocol and Bluetooth networks are referred to as piconets. A piconet consists of a master and up to seven slaves. In the context of connecting a Bluetooth keyboard, the host (SoC or application processor) will be the master and the keyboard the slave. See Figure 11.2 for a pictorial representation. In the figure "M" indicates a master and "S" indicates a slave. The diagram on the left depicts a single slave connected to a master. The one on the right depicts the maximum of seven slaves connected to a master.

The master coordinates communication throughout the piconet. It can send data to any of its slaves and request data from them as well. Slaves are only

---

[2]In addition, Wi-Fi also makes use of the 5 GHz frequency band for some variants.

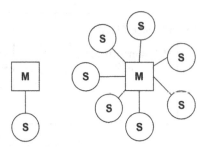

■ **FIGURE 11.2** An illustration of Bluetooth piconets.

allowed to transmit to and receive from their master. They cannot talk to other slaves in the piconet.

Every single Bluetooth device has a unique 48-bit address, commonly abbreviated BD_ADDR. This will usually be presented in the form of a 12-digit hexadecimal value. The most-significant half (24 bits) of the address is an organization unique identifier, which identifies the manufacturer. The lower 24 bits are the more unique part of the address.

Before two Bluetooth devices can begin to communicate, a connection has to be established between them, using the process outlined below:

1. *Inquiry*: If the two Bluetooth devices do not know anything about each other, one of the devices must run an inquiry to find out that the other even exists. When a device receives an inquiry request, it responds with an address and possibly its name and other information such as the kind of Bluetooth profiles[3] it supports.
2. *Paging*: Paging is the process of establishing a connection between two devices based on the addresses found out during the inquiry phase.
3. *Connection*: After completing the paging state, the devices will be put into connection state. While in connection state, a device can be actively transmitting or receiving, or can be in a low-power state.

Table 11.3 shows the various power modes for Bluetooth devices.

The act of establishing a connection between two Bluetooth devices is known as *pairing*. Once two Bluetooth devices are paired together, they will automatically establish a connection when they are within range. This is one of the most convenient features of Bluetooth.

---

[3]A Bluetooth profile is a subspecification that applies to devices that support the said profile. Some example profiles: Human Interface Device Profile (for keyboards) and Headset Profile (for headsets). A single device can support multiple Bluetooth profiles.

Table 11.3 Bluetooth Device Power Modes	
Active mode	The active mode is the regular connected mode, where the device is actively transmitting or receiving data
Sniff mode	Sniff is a power-saving mode, where the device is less active. The device will sleep and only listen for transmissions at a set interval
Hold mode	Hold mode is a temporary, power-saving mode where a device sleeps for a defined period and then returns back to active mode when that interval has passed. The master can command a slave device to hold
Park mode	Park is the deepest of sleep modes. A master can command a slave to "park," and that slave will become inactive until the master tells it to wake back up

Waking a computer

In the context of a keyboard, one of the useful features it can support is the ability to wake a computer from a low-power (sleep) state when a key is pressed. USB keyboards offer this functionality by default.

Let us pause to think what is required for a keyboard to support waking a computer from low-power modes. It just requires that the keyboard be able to communicate with the computer. This in turn requires that the interface used by the keyboard be powered on when the computer is in its low-power/sleep state.

So, whether a Bluetooth keyboard is able to wake a computer from sleep depends on the platform designer. If the Bluetooth interface is not powered down completely during system sleep, waking through a Bluetooth keyboard is possible.

MOUSE

The mouse is connected through almost the same interfaces as a keyboard: USB, Bluetooth, and PS/2. Since these three interfaces have already been discussed in the keyboard section, they will not be discussed here.

In addition, the RS-232 (using the DE9 connector, also incorrectly referred to as DB9 sometimes) interface was also used for connecting computer mice early on. Though this interface is now not used for connecting mice, brief details on it are presented here for completeness.

RS232 (DE9 connector)

RS232 is a standard for serial communication of data. An RS232 connector is shown in Figure 11.3.

Pin	Description
1	Carrier detect
2	Received data
3	Transmitted data
4	Data terminal ready
5	Ground
6	Data set ready
7	Request to send
8	Clear to send
9	Ring indicator

■ **FIGURE 11.3** RS232 pin descriptions.

RS232 interface details follow typical UART[4] interface specifications, a line each for transmitting and receiving data, an Request-to-Send line to request data, and a Clear to Send line to signal to the other device that it can start sending data. Please refer to earlier chapters for more details on the RS232 and UART interfaces.

REMOTE CONTROL

A remote control (or controller), as the name suggests, is a device that is used to operate, control, or command an associated device wirelessly. The range of a remote control is typically less than 10 m. In some cases, remote controllers are required (and able) to communicate over hundreds of meters, as in the case of RC airplanes.

The short-distance remote controls use infrared (IR) as the communication medium, while long-distance ones have to use radio waves. This is because IR has limited strength to go beyond a certain distance. IR remotes also need line of sight to work, which is not the case for radio-based waves. Some recent remote controllers operate using Bluetooth. It must be noted that these three mediums—IR, radio, and Bluetooth—are the most common ones that we will discussing in a bit of detail. However, there are many other protocols that can be used for remote communication. The only criterion is that the controller and controllee should be able to communicate with each other.

---

[4]See Chapter 10, for more details on the UART interface.

Operation

The controller and controllee have to agree on the format and meaning of control codes beforehand. Each specific code means a different action/control to the controllee. The codes are transmitted to the controllee by controller as required, via wireless interface, as illustrated in Figure 11.4.

IR remote controls

IR is the prevalent remote-control technology that is used in home-theater applications. An IR remote control uses light to carry signals between a remote control (controller) and the device it is controlling (controllee). IR remote controls have a typical range of 10 m. IR light is in the invisible portion of the electromagnetic spectrum.

In an IR-based remote control, the controller (the transmitter) sends out pulses of IR light that represent specific binary codes. These binary codes correspond to specific commands that need to be carried out, such as Power On/Off and Volume Up/Down. The receiver in the controllee device (typically, TV, stereo, or other device where IR remotes are primarily used) decodes the pulses of light into the binary digits. The stream of binary digits is sent to the device's microprocessor, which in turn interprets and carries out the corresponding command/operation.

The basic parts involved in sending an IR signal are

- buttons,
- integrated circuit,
- button contacts, and
- LED.

Low wavelength waves (such as IR) cannot go around obstacles. They require line of sight to work. High wavelength waves (such as radio) have higher wavelengths, and can go around obstacles.

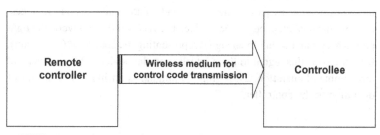

■ **FIGURE 11.4** Logical connection of a remote control.

Mechanism of work

There is a series of activities that happen in carrying out the remote-control operation. The process can be summarized as follows:

1. There are a number of buttons on the remote control. There is a specific code designated for each of them. The code is already agreed between the remote controller and controllee.
2. When we push a specific button, say "volume down" on the remote control, the mechanical arrangement detects the press and the integrated circuit on the remote sends the binary "volume down" command to the LED at the front of the remote.
3. The LED will send out a series of light pulses that correspond to the binary sequence sent to it.

For example, the following is one instance of remote-control codes from the Sony Control-S protocol used in the Sony IR remote controller.

The Sony Control-S protocol includes the following 7-bit binary commands (source: "howstuffworks.com: how remote controls work"):

- Button 1 = 000 0000
- Button 2 = 000 0001
- Button 3 = 000 0010
- Button 4 = 000 0011
- Channel Up = 001 0001
- Channel Down = 001 0001
- Power On = 001 0101
- Power Off = 010 1111
- Volume Up = 001 0010
- Volume Down = 001 0011

The actual remote signal transmitted is not just the code for the command. It includes several other pieces of information, including command start indicator, device address (optional), and command stop indication. After the IR receiver on the device receives the signal from the controller, it verifies the address code as part of the command. If the address code matches, the command is intended for the device. The receiver then converts the light pulses back into the electrical signal representing the command (bit stream). It then passes this signal to the controller on the device, which carries out the command/operation. The "stop" command is an indication of end of operation to the controller.

Pros and cons

IR remotes are the most popular in living rooms today. A number of characteristics contribute to that. First, IR remotes have a limited range of only about 30 ft (10 m), and they require line of sight. This means the IR signal won't transmit through walls or around corners; we need a straight line to the device we are controlling. This also means that people from the neighboring house cannot control devices in our house. But IR light is so ubiquitous that interference can be a problem with IR remotes; for example, sunlight and fluorescent bulbs are the most common source of IR interference. To avoid interference caused by other sources of IR light, the IR receiver on a TV only responds to a particular wavelength of IR light, usually 980 nm—other frequencies are filtered out. Since sunlight can contain IR light of 980-nm wavelength, it may still create interference. To alleviate the problem, the light from an IR remote control is modulated to a frequency not present in sunlight, and the receiver only responds to 980-nm light modulated to that frequency.

While IR remotes are the dominant technology in home-theater applications, there are other specific remote applications that use radio waves instead of light waves. A garage door opener, for instance, makes use of an RF remote. In the next section we will talk about radio-based remote controls.

Radio-based remote controls

Instead of sending out light signals for transmitting the command code, an RF remote transmits radio waves that correspond to the binary command for the command to be carried out. The rest of the process and philosophy remains the same as in an IR remote: meaning there are codes defined for all the commands/buttons, the remote controller sends out the sequence representing the command to be carried out, the receiver on the controllee device receives the sequence and passes that on to the microcontroller on the device. The microcontroller in turn will carry out the operation.

The problem, however, with an RF remote is the potential interference, due to the number of radio signals flying through the air all the time. Cell phones, walkie-talkies, Wi-Fi setups, and cordless phones are all transmitting radio signals at varying frequencies. RF remotes address the interference issue by transmitting at a specific range (band) of radio frequencies and by embedding digital address codes in the radio signal. With the address embedded in the code, the radio receiver on the controllee device can decide whether the command is intended for the device and it has to respond to the command or ignore it.

The key reason to use radio-frequency remotes over IR remotes is their longer range: greater than either IR or Bluetooth; and radio signals can go through walls.

Types of radio control

Broadly speaking, there are two types of radio control, known as single channel and multichannel.

■ Single-channel radio control is effectively an on-off switch operated at a distance by radio wave. The controller unit consists of a low-powered radio transmitter, and the controllee needs a radio receiver and a relay (depending on the usage). The relay is to convert the low-powered, incoming radio signal into a higher-powered electric current. A simple example for using a single-channel control would be to remotely switch a lamp on or off. Single-channel radio control can only switch things completely on or completely off; it cannot turn them up or down by degrees.

■ Multichannel radio control is used to transmit a more complex command pattern to a piece of remote equipment. Instead of just sending a basic on/off signal, it transmits a series of coded analog or digital pulses that are decoded by the receiver and used to produce specific actions. For example, turning a steering wheel on a radio-control transmitter will send a command that includes various parameters of rotation: the direction, degree, and so on. Typically a servo motor in conjunction with the radio receiver is used to carry out the operation. Unlike normal electric motors, servo motors are much more controllable and can be rotated by reasonably precise amounts.

Bluetooth-based remote controls

Bluetooth-based remotes are used for control operations in the smartphone and PC world. The philosophy and operation of these remote controls remain the same as for IR- and radio-based remote controls. The only change is the medium of transmission of control commands. The Bluetooth protocol is explained in detail earlier in this chapter.

SUMMARY

Keyboards and mice are typically interfaced through USB and Bluetooth protocols. USB is the preferred interface because of its widespread availability and ready software support. The Bluetooth interface is preferred as a way of reducing clutter by reducing the number of wires used. Other legacy interfaces such as PS/2 and DIN5 were also discussed briefly in this chapter.

The basic philosophy is the same for IR-, Bluetooth-, and RF-based remote controls. They differ primarily in their range of operation and the media used. While IR remotes have very low range, typically around 10 m, RF-based remote controls have longer ranges, usually in hundreds of meters. The use case dictates the type of remote control to use. For low-range, low-power applications, IR remotes are preferred. For long-range applications, RF remotes are preferred and they consume higher power than IR remotes.

Debug Interfaces

In the earlier chapters, we discussed functional interfaces, the ones used to connect components providing end user functionalities; for example, Bluetooth, Wi-Fi, audio, graphics, and so on. Due to the complexity of the system on chip (SoC) and the integration challenges of components from various different vendors, it is important to have suitable debug interfaces that will allow for quick and efficient debug, determination of root cause, and fix implementation in the design. The scope of debug could be firmware, a software driver, or even the hardware itself. In this chapter we discuss a few of the most common interfaces used for debugging in mobile SoC products, to give an idea of how things work.

DEBUGGER SETUP AND OPERATION

There are two components in the setup: the debugger and the debugged. The debugger and debugged are connected over one of many possible debug interfaces. At the high level, there are two debug methodologies: active debug and passive debug.

- *Passive debugging*: The simplest method is passive debugging, wherein the device being debugged sends out debug messages over a debug interface, and the debugger host captures and displays the messages. These debug messages can possibly be any text or data that helps in identifying the state, control, or data flow of the device being debugged. It is notable that the debugger only receives the messages sent out by the debugged and displays them. The responsibility of spitting out useful messages is on the debugged. The debugger cannot ask for data that it wants, because the communication is one directional. This kind of setup can be visualized as in Figure 12.1.
  This kind of setup is nonintrusive, and the debugged and debugger do not require major infrastructure, neither do they have to follow any specific protocol. All that is required is that the debugged initializes and starts writing to the debug interface; the debugger, if connected and available, will configure itself to the baud rate of the debugged and start receiving messages. Because of its simplicity and innocuousness,

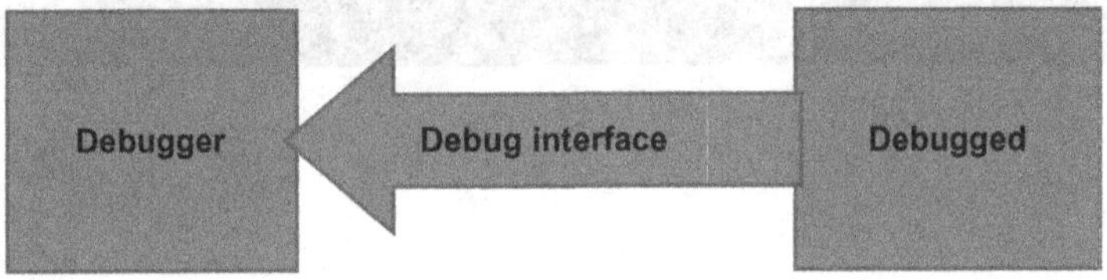

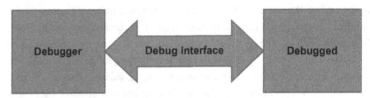

■ **FIGURE 12.1** A view of a debugger operation in passive mode.

■ **FIGURE 12.2** A view of debugger operation in active mode.

this kind of setup is used during early boot flow of system, devices, and CPUs.

■ *Active debugging*: The second scenario (see Figure 12.2) is when the device being debugged is actively involved in inquiring, and/or changing the state, control, or data flow of the device being debugged. In this case, in addition to the messages (data/state) that the debugged sends out, the debugger can ask for additional data needed for debug. Not only can the debugger ask for additional data and state information, but also the debugger can change the control flow (halt the execution of the debugged, or change the program counter) or change the data, thereby triggering a control flow change, and so on. The debugger sends a command over the debug interface to the debugged, and the debugged responds to the commands. Since there is back-and-forth communication of command and data, a client is needed on both sides that listens from the other end and responds appropriately. Due to the need for clients, this mechanism can be employed only after the basic infrastructure is up and the clients are up and working.

Now there are multiple options and possibilities for the debug interface. Some of these are software-only debug; others allow probing and debugging the underlying hardware (SoC). Some of the most common examples of the debug interfaces that we will talk about are Universal Asynchronous Receiver Transmitter (UART), Universal Serial Bus (USB), Joint Test Action Group (JTAG), and NEXUS.

UART

As discussed in earlier chapters, UART is the oldest, simplest, yet most common serial interface in use today. The UART supports limited bandwidth when compared with some of the modern interfaces; however, this low bandwidth characteristic does not prohibit UART's usage as a debug interface.

The UART interface is typically used in both active and passive debug methodologies.

- *UART usage for passive debugging*: Since the UART is a simple protocol and initialization of UART is quite easy, the UART is used as a preferred interface for passive debugging employed during early boot flow of system, devices, and CPUs. For example: BIOS uses UART for debug messages in passive debug mode.
- *UART usage for active debugging*: UART is also commonly used in active debug setups. In this case the command and data travel through UART as the debug interface. The debugger sends commands over the UART to the debugged, and the debugged responds to the commands. Since there is back-and-forth communication of commands and data, a client is needed on both the sides that listens from the other end and responds appropriately. A typical example of active debugging setup is Windbg connection over UART. Windbg, a short form for Windows debugger, is one of the most commonly used debug tools for the Windows platform. Windbg is used for both kernel mode (driver) and user mode (application) debug.

Due to slower speed, however, the UART is losing out to some of the speedier interfaces like USB.

USB

Although UART continues to be the choice for early boot, passive debug mode, USB is rapidly replacing UART for active debug mode (Figure 12.3).

■ **FIGURE 12.3** A view of USB 2.0/3.0 debugger operation in active mode.

For USB 2.0 to be useful as debug interface, the USB 2.0 (also known as EHCI or Enhanced Host Controller Interface for USB) controller, has to implement the debug port capability. The debug port capability is an optional capability of EHCI controllers. The debug port provides a mode of operation that requires neither RAM nor a full USB stack. A handful of registers in the EHCI controller Peripheral Component interconnect (PCI) configuration and BAR address space are used for all the communication. All three transfer types are supported (OUT/SETUP/IN), but transfers can only be a maximum of 8 bytes and only one specific physical USB port can be used. A Debug Class compliant device is the only supported USB function that can be communicated with.

Even though the USB 2.0 was usable as debug interface, it had limitations, and the debug setup with USB 2.0 was complex. Here are a few of the limitations on USB 2.0 for debug (taking Windows as an example here):

1. Only first port could be used for debug.
2. Special hardware device was needed to connect the host and target.
3. Setup was unstable; in some cases, a reboot with the device attached would hang the system during BIOS POST. This precluded the debug of early boot flow.

USB 3.0 was designed from the ground up to provide operating system kernel debug capability for Windows.

The prerequisites for USB 3.0 debugging are

- A host system with an xHCI (USB 3.0) host controller. The USB 3.0 ports on the host system do *not* need USB 3.0 debug support—only the target system must have that.
- A target system with an xHCI (USB 3.0) host controller that supports debugging.
- A USB 3.0 (A-A) crossover cable.

JOINT TEST ACTION GROUP

JTAG is the common name used for the IEEE 1149.1 Standard Test Access Port and Boundary-Scan Architecture. It was originally devised and applied for testing printed circuit boards. It is still widely used for its original application. In addition to that, JTAG has evolved as a key IC debug port. It is also prominently used in system programming interface (such as for programming field programmable gate array [FPGA]).

Evolution

JTAG was conceived to address difficulties in testing circuit boards after manufacturing. At the time, multilayer boards and nonlead-frame ICs were becoming standard, and connections were being made between ICs that were not available to probes. This was aggravated due to modern packaging technologies such as ball grid array (BGA) and chip-scale packaging limiting and in some cases eliminating physical access to pins.

The majority of manufacturing and field faults in circuit boards were due to solder joints on the boards, imperfections in board connections, and so forth. JTAG was meant to provide a pins-out view from one IC pad to another so all these faults could be discovered.

JTAG fundamentals

JTAG solved this challenge by placing cells between the external connections and the internal logic of the device, as seen in Figure 12.4. With the cells configured as a shift register, JTAG can be used to set and retrieve the values of pins (and the nets connected to them) without direct physical access.

There is also an option to sample the data values as they pass between the core logic and the pins during the normal operation of the device.

The JTAG protocol adds four extra pins to device:

- TDI (Test Data In) to input data to the device,
- TDO (Test Data Out) to output data from the device,
- TMS (Test Mode Select) to control what should be done with the data and a clock signal, and
- TCK (Test Clock) to synchronize everything.

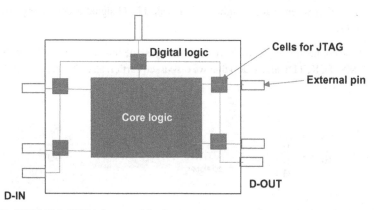

■ **FIGURE 12.4** JTAG interface connectivity diagram.

In addition there is an optional signal, TRST, an active-low reset to the test logic. The TRST is usually asynchronous to other signals.

So, the JTAG interface should have at least these four pins. However, there is a reduced pin JTAG interface. Reduced pin JTAG uses only two wires, a clock wire and a data wire. This is defined as part of the IEEE 1149.7 standard. The connector pins are TMSC (Test Serial Data) and TCK (Test Clock). It is also possible to connect more than one module or chip together to form a chain.

Communication model

In JTAG, devices expose one or more test access ports (TAPs). Each of these TAPs can be individual chips or might be modules inside one chip. A daisy chain of TAPs is called a *scan chain*. Scan chains can be arbitrarily long.

To use JTAG, a host is connected to the target's JTAG signals (TMS, TCK, TDI, TDO, and so on) through some kind of JTAG adapter. The adapter connects to the host using some interface such as USB, PCI, or Ethernet. Figure 12.5 illustrates a setup using the JTAG interface.

Connectors

There are no official standards for JTAG adapter physical connectors. The connector depends on the tool that will be using the JTAG interface. Development boards usually include a header to support preferred development tools; in some cases it may include multiple headers to support multiple such tools. For example, a microcontroller, FPGA, and ARM application processor will rarely share tools, so a development board using all of those components might have three or more headers. Production boards may omit the headers; or when space is tight, just provide JTAG signal access using test points.

These connectors typically have more than just the four standardized signals (TMS, TCK, TDI, and TDO) that we discussed earlier. Usually, the optional,

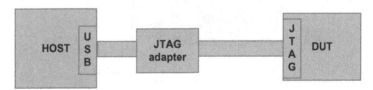

■ **FIGURE 12.5** Illustration of a debug setup with JTAG.

reset signals are also provided: one or both of TRST (TAP reset) and SRST (system reset). The connector usually provides the board-under-test's logic supply voltage so that the JTAG adapters will use the appropriate logic levels.

The JTAG architecture is simple yet powerful for debug at the lowest level of the logic in SoC. Using JTAG, the debugger can inspect the state of all the TAPs in the target (scan chain). Since the JTAG is available from the very beginning of the system boot-up, JTAG can be used to diagnose problems at any time. JTAG is used for debugging both software and hardware problems, equally effectively. It must be noted that, at times, the interface designer and developers may choose to implement only a part of the specification.

NEXUS

Nexus has established itself as a prominent debugging interface for embedded systems. Nexus is formally known as IEEE-ISTO 5001-2003. Nexus attempts to create rich debug feature set with minimum pins and die area. In addition, Nexus standard does not want to bind to specific processor or architecture; instead it wants to be both processor and architecture independent. The standard supports multicore and multiprocessor systems. Nexus standardizes the connectors used for connecting the debug tool to the target or system under test. The data transfer happens using packet-based protocol. The actual physical medium of data transfer can be any other protocol; for example, JTAG or depending on the required transfer speed it can be an auxiliary port meeting the speed requirement.

Nexus categorizes the debug capabilities in the following major categories:

- *Run time control*: This refers to controlling the operation of processor.
- *Memory access*: This refers to the capability of accessing memory while the processor is running.
- *Breakpoints*: This refers to capability of breaking the execution of the processor on specific events.
- *Event tracing*: This corresponds to the capability to generate program execution trace, data access trace, ownership trace (with respect to the process/tasks), and so on.
- *Memory substitution and port replacement*: This feature allows memory or port accesses to be emulated over the auxiliary Nexus port.
- *Data acquisition*: Rapid prototyping may require rapid transfer of large amounts of data via the auxiliary port to the debug tools. It uses a more efficient protocol than that used in data trace. It also helps calibration in automotive applications.

Nexus is a scalable standard and allows for multiple levels of functionality through multiple classes of compliance.

- *Class 1*: Supports runtime control capability using the JTAG interface.
- *Class 2*: Adds ownership trace and program trace to Class 1 functionality.
- *Class 3*: Adds data write trace and memory access capability to Class 2.
- *Class 4*: Adds memory substitution (fetching or reading data over the Nexus auxiliary port) capability to Class 4.

Overview of Intel SoC: Baytrail

Up until now, all our discussion on system on chip (SoC) components and block diagrams has been generic. In this appendix we refer to a real SoC-based platform: Intel's Baytrail reference platform based on Valleyview SoC. The product using Valleyview SoC has already been a huge success in the market and has got Intel a place in the mobile and tablet market. One of the key differentiators for Intel SoC-based products is the legacy support while using the Microsoft Windows operating system, meaning, due to the x86 architecture, all the legacy software that people have been using on PC systems would continue to seamlessly work on the Intel SoC-based platforms. That's a huge advantage for users.

Now, talking about the significance of reference platforms and designs, the way things work is: Intel or any SoC vendor develops a reference design, which has an SoC, board, add-in cards (external devices), BIOS, software drivers, and applications required for feature enabling. The reference is used by the SoC vendor to integrate, validate, and demonstrate the platform capability. Once done, the OEMs/ODMs will take the reference design, tweak the design for value-add components and features, integrate and validate their platform, and sell that in the market. Using the reference as a base and building on top of that is the logical way to efficient time to market: Since the reference design is already known to be working, it's less work and faster to just add a few things and get the design ready for market. However, it is not mandatory that the OEMs should use the reference as a base. At times, the OEMs/ODMs may want to do the platform design from scratch, and they actually do that when needed. The only problem with the design from scratch approach is the additional effort that needs to be put in design, validation, and debug of the platforms.

BAYTRAIL REFERENCE VALIDATION PLATFORM BLOCK DIAGRAM

Figure 3.1, in Chapter 3, depicts the Baytrail connectivity diagram. As can be seen in the diagram, the SoC sits in the middle of the platform (logically), and then there are interfaces coming out of the SoC. These interfaces will connect the end devices that are outside of the SoC. These end devices sit on the motherboard (PCB).

Let's have a look at the components on the board: To start with, the PMIC AIC is responsible for power management and delivery to the SoC. The Wi-Fi chip on the Baytrail reference platform uses the SDIO interface. The WiFi chip is a combo module; it also has an integrated Bluetooth module. The Bluetooth part of the module is connected over UART. The other UART controller is used for interfacing the GPS module. I^2S interfaces are used for audio routing. One I^2S interface is used for audio data transmission to an onboard speaker, the other one to Bluetooth for audio over a Bluetooth usage scenario. The WWAN module is attached over the USB interface, and HSIC is not used. ULPI (ultra low power USB Phy) is used for connectivity of generic USB devices. DDR3L DRAM is connected on memory channel 0, and channel 1 is not used. There are three display ports: one HDMI, one eDP, and one MIPI display. These display ports can be used as per usage requirements. The camera sensor devices are connected on a CSI interface. The eMMC and micro SD cards are supported over the SD interface. The JTAG interface is used to connect debug connector XDP. It is not shown in the diagram, but it should be understood that GPIOs are used across almost all the components for module power/reset control and interrupts from the add-in module to the host processor.

The internal blocks of the Valleyview SoC can be seen in Figure 3.2, in Chapter 3.

As can be seen in the SoC block diagram, there are multiple I^2C, UART, and SPI controllers. It has USB (2.0, 3.0, and HSIC) controllers. The SoC also supports PCIe interfaces. It uses SD and eMMC and SATA for storage, CSI-2 for camera, and MIPI-DSI, eDP, and HDMI for display. Multiple I^2S controllers are supported for streaming audio over the onboard speaker, Bluetooth (audio over Bluetooth usage), and WWAN interfaces (voice call). Both controllers for LPDDR2/3 and DDR3 are part of the SoC for DRAM interfacing. Apart from these interfaces for external interfacing, the CPU, graphics, imaging processor, audio processor, security engine, and so on are all part of the SoC. The whole feature set of the SoC is summarized in Table A.1. In terms of the key power and performance indicators, Table A.2 provides the summary. Please note that Tables A.1 and A.2 provide indicative figure only and are not meant to be reference.

In this appendix we looked at Intel's Baytrail SoC-based reference platform, which was key to getting Intel a foot in the mobile and tablet market. We discussed the internal and external view of the Baytrail reference platform. We also discussed the key aspects of this SoC, the feature set, and key

Table A.1 Valleyview SoC Key Feature Set

Feature	Comment/Measurement
CPU core	Silvermont
Number of CPU cores	4
Memory	LPDDR2/3, DDR3L
Memory speeds	800, 1067, 1333 MT/s
Memory bandwidth	12.8-21.3 GB/s
Graphics	GenLC (Gen 7) D 4 EU (DX11 support)
Display	2 display pipe, 8 lane MIPI DSI DPHY, eDP/DP/HDMI/VGA
Imaging	Silicon Hive ISP, MIPI CSI DPHY—5 lanes
Audio	Tensilica 5.1, Azalia HD audio
USB2	USB2×4
USB HSIC	USB HSIC×2
USB OTG	USB3 OTG
USB3 host	USB3 host ×1
eMMC	eMMC 4.4.1
Security engine	Security engine with NFC
SATA	SATA2×2
PCIe	PCIe2×4
I^2S/SSP	I^2S×3
SPI	SPI×3
I^2C	I^2C×7 3.4 MHz
MIPI HIS	MIPI HIS ×1
SDIO	SDIO3.0 ×2
HSUART	HSUART×2
Package	25×27, BGA
Die size	~100 mm square
Platform interface	LPC, SVID, and GPIO 1.8 V
Power management	S0ix power ~20 mW

Table A.2 Key Power and Performance Indicators

Serial#	Feature	Measurement/Comment
1	Video battery life	>10 h
2	Standby battery life	~30 days
3	Z height	~8 mm or less
4	Skin temp. (over ambient)	~+15 °C for plastic skin and ~12-13 °C for metal skin
5	CPU perf.	1.8×(ST), 4.7×(MT)
6	Threads/cores	4/4
7	Process	22 nm
8	Graphics core	Intel Gen7
9	Windows Gfx capability	DX11

power, and performance indicators. This SoC got rave reviews from analysts for its power efficiency and is the base Intel can build on in the mobile and tablet market. One of the key benefits that Intel SoC-based system users have is that their mobile and tablets can double as devices for other things that typically needed a PC system. All the applications that they used on PC can be used on these tiny mobile/tablet devices. The advantage Intel enjoys is the large software ecosystem (both in terms of system and applications) that exists and is fully compatible with Intel's SoC platforms.

Industry Consortiums

Systems today are called "open systems," meaning different components for the system come from different vendors and players. There are many vendors/players involved in getting the whole stack up to create a system. For example, talking about a mobile infotainment device for an example, there will be SoC vendors, the platform designers, manufacturers, a display designer and manufacturer, touch controller provider, Wi-Fi chip provider, a software developer for all these hardware components, and so on. So, fundamentally, there is a complete ecosystem involved in making a system. Now, since there are so many players, it is imperative for these players to come together and standardize the pieces so it will be easier for everybody to work together in making a system. For this reason there are consortia or special interest groups formed with representation of all these different players (design, validation, manufacturing, software, and so on).

In the following section we briefly describe the major bodies involved in standardization today. We only talk about the prominent bodies in the domain of mobile devices and technology. The list is not intended to be complete. It is just an attempt to provide an overview of the major standards bodies.

1. *VESA*: VESA, or the Video Electronics Standards Association, is an international standards body for computer graphics. It was formed in 1988 by NEC Home Electronics and eight video display adapter manufacturers. The other display adapter manufacturers that were part of the association were: ATI Technologies, Genoa Systems, Orchid Technology, Renaissance GRX, STB Systems, Tecmar, Video 7, and Western Digital/Paradise Systems. VESA's initial goal was to produce a standard for 800×600 SVGA resolution video displays. Thereafter, VESA has created and published a number of standards. The primary area of focus for VESA remains the function of video peripherals in personal computers.
2. *MIPI*: MIPI is the commonly used name for the MIPI Alliance. MIPI is a global, open membership organization. The focus area of MIPI is developing interface specifications relating to the mobile devices.

The alliance was founded in 2003. The founding members were ARM, Intel, Nokia, Samsung, STMicroelectronics, and Texas Instruments. As of today, the alliance has more than 250 member companies worldwide, and around 12 active working groups. The alliance has delivered more than 45 specifications since then. MIPI is one of the most powerful alliances today as far as specifications in mobile device area are concerned.

3. *HDMI*: More formally known as HDMI Licensing, LLC, HDMI was founded by Hitachi, Panasonic, Philips, Silicon Image, Sony, Thomson (RCA), and Toshiba and is their licensing vehicle for the HDMI standard.

4. *JEDEC*: The JEDEC Solid State Technology Association was earlier known as the Joint Electron Device Engineering Council (JEDEC). It is an independent semiconductor engineering trade organization and standardization body. JEDEC is a global leader in developing open standards for the microelectronics industry. It has more than 3000 volunteers representing nearly 300 member companies across the world.

5. *IEEE-SA*: The Institute of Electrical and Electronics Engineers Standards Association (IEEE-SA) is a part of IEEE. The focus area of IEEE-SA is to develop standards in a broad range of industries such as power and energy, biomedical and healthcare, information technology, telecommunication, transportation, nanotechnology, information assurance, and so on. IEEE-SA is one of the oldest bodies focused on standardization. IEEE-SA is not a body formally authorized by any government. It is rather a community that focuses on standardization, and technical experts from all over the world participate in the development of IEEE standards. This is in contrast to the formally recognized international standards organizations (e.g., ISO, IEC, ITU, CEN). These formally recognized bodies are federations of national standards bodies (American ANSI, German DIN, Japanese JISC, and so on). The standards developed by IEEE-SA have names like IEEE 802.11, for example, IEEE 802.11 constitutes the wireless networking technology also known as Wi-Fi.

6. *PCI-SIG*: The PCI-SIG or Peripheral Component Interconnect Special Interest Group is an electronics industry consortium. It is responsible for specifying the base PCI and its derivative busses. The PCI Special Interest Group was formed in 1992, in Beaverton, OR. Initially it started as a "compliance program" to help computer manufacturers implement Intel's PCI specification. Later, in year 2000, the organization became a nonprofit corporation, officially named PCI-SIG. PCI-SIG has more than 800 member companies involved in developing products based on its specifications. It has produced the PCI, PCI-X, and PCI Express specifications and their multiple versions.

7. *USB-IF*: The USB Implementers' Forum (USB-IF) is a nonprofit organization. The forum is responsible for promoting and supporting the Universal Serial Bus. Its objective is to promote and market USB, Wireless USB, USB On-The-Go, maintain the specifications, as well as running the compliance program. It was formed by the group of companies that developed USB, in 1995. The key founding members were Apple Computer, Hewlett-Packard, NEC, Microsoft, Intel, and Agere Systems. It currently has four working committees; namely, the Device Working Group, Compliance Committee, Marketing Committee, and On-The-Go Working Group.

8. *IEC*: The International Electrotechnical Commission (IEC) is a nonprofit, nongovernmental international standards organization. The focus area of IEC is to prepare and publish international standards for electrical, electronic, and related technologies, which are collectively known as "electrotechnology."

9. *ISO*: The International Organization for Standardization, or ISO, is an international standard-setting body. It has representatives from various national standards organizations. The body promotes worldwide proprietary, industrial, and commercial standards. It was founded on February 23, 1947, and is headquartered in Geneva. It has more than 164 countries as members as of today. Additionally, ISO has formed joint committees with the International Electrotechnical Commission (IEC) to develop standards in the areas of electrical, electronic, and related technologies. The joint committees have two key parts: ISO/IEC JTC 1 and ISO/IEC JTC 2. ISO/IEC JTC 1 looks after standards in information technology domain, while ISO/IEC JTC 2 is responsible for standardization in the field of energy efficiency and renewable energy sources.

10. *ITU*: The International Telecommunication Union (ITU) was earlier known as the International Telegraph Union. It is a specialized agency of the United Nations (UN), and it is responsible for issues in the area of information and communication technologies. The ITU coordinates the use of global resources like radio spectrum and promotes international cooperation in assigning satellite orbits. Additionally it works to improve telecommunication infrastructure worldwide and assists in the development and coordination of worldwide technical standards.

11. *CEN*: The European Committee for Standardization (CEN) is a nonprofit organization. The CEN was founded in 1961. CEN is officially recognized as a European standards body by the European Union; the other official European standards bodies are the European Committee for Electrotechnical Standardization (CENELEC) and the European Telecommunications Standards Institute (ETSI).

USB 3.0

USB 3.0, the latest version of USB (Universal Serial Bus), provides better speed and more efficient power management than USB 2.0. USB 3.0 is backward-compatible with USB 2.0 devices; however, data transfer speeds are limited to USB 2.0 levels when these devices interoperate.

Unlike the change from USB 1.0 to USB 2.0, USB 3.0 brings actual physical differences to the connectors. The flat USB Type-A plug (that goes into the computer) looks the same, but inside is an extra set of connectors; the edge of the plug is colored blue to indicate that it is USB 3.0 (see Figure C.1).

On the other end of the cable, the Type B plug (that goes into the USB device) actually looks different because the extra set of connectors and the Type B plug (see Figures C.2 and C.3) could not be accommodated in the smaller USB 2.0 plug.

As a result, we cannot plug a USB 3.0 cable into a USB 2.0 device. However, we can plug USB 3.0 devices and cables into a computer having a USB 2.0 port; however, we will not get the speed advantage. The reason for the new connector is that the USB 3.0 cable contains nine wires, which is four more than a USB 2.0 cable; eight carry data, and one is used as a ground. Despite the increase in wires, however, the cables should almost be of the same width as the USB 2.0 cables. There will be a big difference in performance, however.

MOTIVATION FOR USB 3.0

USB 2.0 was a hugely popular interface for connecting devices to computers. Building off the success of USB 2.0, the third iteration of the USB aims to tackle three fundamental pillars of interface design: speed, power, and interoperability.

Speed

Demand for faster speed became stronger. Compared to the paltry 12 Mbps speeds of USB 1.1 and the moderate 450 Mbps of "High-Speed" USB 2.0, the "SuperSpeed" interface of USB 3.0 tries to live up to its name with a theoretical 5.0 Gbps.

■ **FIGURE C.1** USB 3-A connector.

■ **FIGURE C.2** USB 3-B connector.

■ **FIGURE C.3** USB 3.0 Micro AB.

Power

With an ever-expanding list of accessories and portable devices, bus-powered hardware has long been pushing the limits of what USB 2.0 could handle. First, the 3.0 specification allows up to 80% higher power consumption for devices running at SuperSpeed, and second, USB 3.0 includes an enhanced version of the USB-B connector called Powered-B, which allows USB accessories to draw power from peripheral devices as well as hosts.

Interoperability and cross configuration

New features have been implemented in USB 3.0 to establish a robust ecosystem of USB devices' cross-communication between hardware devices. Similar to FireWire and Ethernet specifications, the new SuperSpeed USB includes an established method of host-to-host communication through a crossover USB-A to USB-A cable. Additionally, USB 3.0 builds on the "USB OTG" principles of allowing portable devices like phones to act as either a USB device or a USB host.

PROTOCOL OVERVIEW

USB 3.0 is a physical SuperSpeed bus combined in parallel with a physical USB 2.0 bus. From the architecture perspective it has similar components to USB 2.0, that is:

- USB 3.0 interconnect
- USB 3.0 device
- USB 3.0 host

Figure C.4 shows the USB 3.0 architecture.

The USB 3.0 interconnect is defined as the mechanism with which USB 3.0 and USB 2.0 devices connect to and communicate with the host. The architecture of the USB 3.0 interconnect is inherited from USB 2.0. However, many of the architectural elements are augmented to support the dual bus architecture of USB 3.0, which is meant to provide backward/forward compatibility with USB 2.0. The fundamentals and the structural topology of USB 3.0 are the same as USB 2.0. That means: USB 3.0 also has a tiered-star topology. The topology starts with a single host at the very first tier. It has hubs at lower tiers to provide connectivity to the end devices. Due to the dual bus architecture the USB 3.0 can support backward and forward compatibility, which means that both USB 3.0 and USB 2.0 devices can be plugged into a USB 3.0 bus.

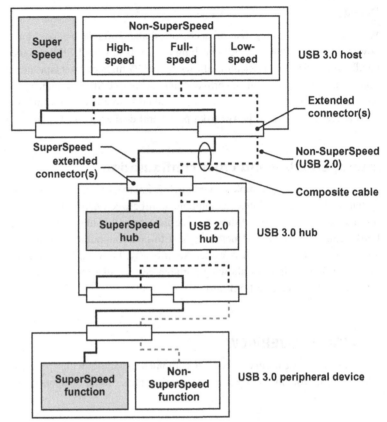

Note: Simultaneous operation of SuperSpeed and non-SuperSpeed modes is not allowed for peripheral devices.

■ **FIGURE C.4** USB 3.0 architecture.

Similarly, USB 3.0 devices can be connected to a USB 2.0 bus. The mechanical and electrical backward/forwards compatibility for USB 3.0 is again possible due to the composite cable and associated connector assemblies. The composite cable and associated connector assemblies basically form the dual bus architecture. In other words, USB 3.0 devices have both Super-Speed and non-SuperSpeed bus interfaces. The non-SuperSpeed part of the interface is responsible for providing the backward compatibility with USB 2.0 interface. Similarly, USB 3.0 hosts include both SuperSpeed and non-SuperSpeed bus interfaces. The non-SuperSpeed part of the interface is responsible for interfacing with USB 2.0 devices if plugged. Both the Super-Speed and non-SuperSpeed parts of the bus interface are parallel buses that may be active simultaneously. USB 3.0 hubs, again similar to their USB 2.0

counterparts, are a specific class of USB device. The purpose of USB hubs is to provide additional connection points to the bus if more than what is provided by the host is required.

The physical interface of USB 3.0 also has USB 2.0 electrical, mechanical, in addition to the expected SuperSpeed physical interface for the buses. Therefore, USB 3.0 cables have eight primary conductors: three twisted signal pairs for USB data paths and a power pair.

Figure C.5 shows a logical diagram of the USB connectors.

The device discovery, configuration, and other characteristics of USB 3.0 remain the same as USB 2.0. Table C.1 attempts a comparison of various USB characteristics between USB 2.0 and USB 3.0.

SUMMARY

So, as can be seen from the earlier description, USB 3.0, the next-generation USB interface to USB 2.0, is backward-compatible to USB 2.0, and it provides higher speed, better power management, scalability, and interconnectability. Despite the fact that USB 3.0 is backward-compatible to USB 2.0, there are some fundamental differences.

Given USB's prominence in the market today, and given that USB 3.0 is addressing the data rate concerns that new usage and devices were starting to pose with USB 2.0, and improving the power management even further to lower overall power consumption, USB is going to remain the interface of choice for device interfacing.

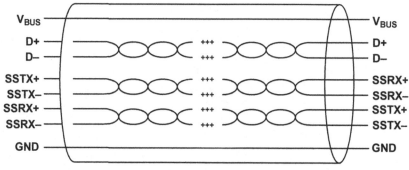

■ **FIGURE C.5** USB 3.0 cable connector.

Table C.1 Comparison of Key Attributes in USB 2.0 and USB 3.0 Interface

Characteristic	USB 2.0	USB 3.0
Standard release	April 2000	November 2008
Data rate	Low-speed (1.5 Mbps), full-speed (12 Mbps), and high-speed (480 Mbps)	SuperSpeed (5.0 Gbps)
Data interface	Half-duplex two-wire differential signaling Unidirectional data flow with negotiated directional bus transitions	Dual-simplex, four-wire differential signaling separate from USB 2.0 signaling Simultaneous bidirectional data flows
Number of wires (within cable)	4	9
Bus transaction protocol	Host directed, polled traffic flow Packet traffic is broadcast to all devices	Host directed, asynchronous traffic flow Packet traffic is explicitly routed
Power management	Port-level suspend with two levels of entry/exit latency Device-level power management	Multilevel link power management supporting idle, sleep, and suspend states. Link, device, and function-level power management
Bus power	Support for low/high bus-powered devices with lower power limits for un-configured and suspended devices	Same as for USB 2.0 with a 50% increase for unconfigured power and an 80% increase for configured power
Port state transition	Port hardware detects connect events. System software uses port commands to transition the port into an enabled state	Port hardware detects connect events and brings the port into operational state ready for SuperSpeed data communication
Maximum cable length	5 m	3 m

USB OTG (On The Go)

USB On-The-Go, also known as USB OTG, is a specification that allows USB devices such as digital audio players, printers, or mobile phones to act as a host. This capability allows some other USB devices like a USB flash drive, digital camera, mouse, or keyboard to be attached to them while allowing them to function normally. This kind of capability enables a number of exciting features; for example, a digital camera can directly be connected to printer (acting as host) for printing pictures, or audio player (acting as host) can connect to cell phone and download songs, etc.

So, fundamentally, use of USB OTG allows these devices to switch back and forth between the roles of host and client devices. For instance, a mobile phone may read from removable media as the host device or behave as a USB Mass Storage Device when connected to a host computer.

MOTIVATION AND PROTOCOL OVERVIEW OF USB OTG

USB, the most common interface today, was originally designed as an interface between PCs and peripherals. By definition, USB communication occurs between a host and a peripheral. The original intent was to place the heavier workload on the PC (host) and to allow USB peripherals to be fairly simple. Accordingly, the USB specification requires that PCs do most of the work, such as, for example, to provide power to peripherals, and to support all defined speeds and data flow types.

As computing resources became less expensive, the line between PCs and other products blurred, and therefore many devices that are not PCs in the classic sense have a need to connect directly to peripherals: printers connect directly with cameras, for example, or mobile phones may need to connect to USB headsets, and so on. So, to enable these kinds of usage scenarios, these devices added the USB host function; however, they needed to function in ways that differ from standard PC USB hosts: although they need to provide host capability for some devices, they are not required to support the full range of USB peripherals. For example, connecting a camera to a printer makes a lot of sense, but the printer manufacturers may not think it is quite as important for the printer to support a USB GPS dongle.

So, these devices acting as USB hosts are defined as Targeted Hosts. A Targeted Host is a USB host that supports a specific, targeted set of peripherals. The developer of each Targeted Host product defines the set of supported peripherals on a Targeted Peripheral List (TPL). A Targeted Host needs to provide only the power, bus speeds, data flow types, and so on, that the peripherals on its TPL require. There are two categories of Targeted Hosts:

1. *Embedded Hosts*: An Embedded Host (EH) product provides Targeted Host functionality over one or more Standard-A or Micro-AB receptacles. EH products may also offer USB peripheral capability, delivered separately via one or more Type-B receptacles.
2. *On-The-Go*: An OTG product is a portable device that uses a single Micro-AB receptacle (and no other USB receptacles) to operate at times as a USB Targeted Host and at times as a USB peripheral. OTG devices will operate as a standard peripheral when connected to a standard USB host.

OTG devices can also be attached to each other. There are protocols defined in the specification to identify the master (host) and slave (peripheral) in such cases. The USB OTG and EH Supplement to the USB 2.0 specification introduced three new communication protocols:

- *Attach Detection Protocol* allows an OTG device, EH, or USB device to determine attachment status in the absence of power on the USB bus.
- *Session Request Protocol* allows both communicating devices to control when the link's power session is active. In standard USB, only the host is capable of doing so. This allows fine control over the power consumption.
- *Host Negotiation Protocol* (HNP) allows the two devices to exchange their host/peripheral roles, provided both are OTG dual-role devices. By using HNP for reversing host/peripheral roles, the USB OTG device is capable of acquiring control of data-transfer scheduling.

The actual protocol and the framework of data transfer on OTG devices remain the same as USB 2.0 (and USB 3.0 for SuperSpeed OTG devices). The OTG devices will behave as standard USB hosts or devices when connected to standard (non-OTG) USB devices.

References

1. http://www.uefi.org
2. http://www.trustedcomputinggroup.org
3. http://www.pcisig.com/home
4. AGP3.0 interface specification
5. ACPI 5.0 specification
6. Baytrail platform External Design specification
7. Computer Organization and Design, Fifth Edition: The Hardware/Software Interface (The Morgan Kaufmann Series in Computer Architecture and Design
8. TCG Specification Architecture Overview version 1.4
9. SD Host controller Simplified specification version 2.00
10. http://www.usb.org
11. http://www.mipi.org
12. Information technology – AT attachment 8 – ATA/ATAPI command set (ATA8-ACS) revision 4a May 21, 2007
13. Intel Low Pin Count Interface Specification_v1.1. Document Number 251289–001
14. https://en.wikipedia.org/wiki/Floyd%E2%80%93Steinberg_dithering
15. https://en.wikipedia.org/wiki/Intel_Display_Power_Saving_Technology
16. Digital Visual Interface DVI Revision 1.0, 02 April 1999
17. VESA Enhanced Extended Display Identification Data Standard (Defines EDID Structure Version 1, Revision 3) Release A, Revision 1 February 9, 2000
18. VESA Notebook Panel Standard Version 1 October 22, 2007
19. VESA DisplayPort Standard Version 1, Revision 1a January 11, 2008
20. VESA Embedded DisplayPort (eDP) Standard Version 1.4 28 February, 2013
21. High-Definition Multimedia Interface Specification Version 1.3a November 10, 2006
22. High-Definition Multimedia Interface Specification Version 1.4a Extraction of 3D Signaling Portion March 4, 2010
23. Display Data Channel Command Interface Standard Version 1.1 October 29, 2004
24. VESA Enhanced Display Data Channel (EDDC) Standard Version 1.2 December 26, 2007
25. DDR2 SDRAM SPECIFICATION JESD79-2F November 2009
26. PC2–5300/PC2-6400 DDR2 SDRAM Unbuffered DIMM Design Specification JEDEC Standard No. 21C
27. MIPI Alliance Specification for Display Serial Interface (DSI^SM) Version 1.3 –
28. MIPI Alliance Specification for D-PHY^SM Version 1.2
29. MIPI Alliance Specification for Serial Low-power Inter-chip Media Bus (SLIMbus®) Version 1.01.01

30. MIPI Alliance Specification for Camera Serial Interface 2 (CSI-2^{SM}) Version 1.00
31. www.wikipedia.org
32. Universal Serial Bus Specification Revision 2
33. Texas instrument AN-452 MICROWIRE Serial Interface
34. Philips Semiconductor I2S specification

Index

Note: Page numbers followed by *f* indicate figures and *t* indicate tables.

Printed in the United States
By Bookmasters